Test and Evaluation Trends and Costs for Aircraft and Guided Weapons

T0308366

Bernard Fox, Michael Boito, John C.Graser, Obaid Younossi

Prepared for the United States Air Force

Approved for public release, distribution unlimited

PROJECT AIR FORCE

The research reported here was sponsored by the United States Air Force under Contract F49642-01-C-0003. Further information may be obtained from the Strategic Planning Division, Directorate of Plans, HqUSAF.

Library of Congress Cataloging-in-Publication Data

Test and evaluation trends and costs for aircraft and guided weapons / Bernard Fox ... [et al.].
    p. cm.
    "MG-109."
    Includes bibliographical references.
    ISBN 0-8330-3540-1 (pbk. : alk. paper)
    1. Airplanes, Military—United States—Testing. 2. Antiaircraft missiles—United States—Testing. 3. Airplanes, Military—United States—Costs. 4. Antiaircraft missiles—United States—Costs. 5. Air-to-surface missiles—Testing. 6. Air-to-surface missiles—Costs. 7. United States. Air Force—Weapons systems—Testing. 8. United States. Air Force—Weapons systems—Costs. I. Fox, Bernard, 1951–

UG1243.T47 2004
358.4'18—dc22

    2004005294

Cover photo courtesy of the U.S. Air Force at www.af.mil.
Photographer: Judson Brohmer

The RAND Corporation is a nonprofit research organization providing objective analysis and effective solutions that address the challenges facing the public and private sectors around the world. RAND's publications do not necessarily reflect the opinions of its research clients and sponsors.

RAND® is a registered trademark.

Published 2004 by the RAND Corporation
1700 Main Street, P.O. Box 2138, Santa Monica, CA 90407-2138
1200 South Hayes Street, Arlington, VA 22202-5050
201 North Craig Street, Suite 202, Pittsburgh, PA 15213-1516
RAND URL: http://www.rand.org/
To order RAND documents or to obtain additional information, contact
Distribution Services: Telephone: (310) 451-7002;
Fax: (310) 451-6915; Email: order@rand.org

Preface

This is one of a series of reports from a RAND Project AIR FORCE project, "The Cost of Future Military Aircraft: Historical Cost Estimating Relationships and Cost Reduction Initiatives." The purpose of the project is to improve the tools used to estimate the costs of future weapon systems. It focuses on how recent technical, management, and government policy changes affect cost.

This monograph examines the effects of changes in the test and evaluation (T&E) process used to evaluate military aircraft and air-launched guided weapons during their development programs. Working from extensive discussions with government and industry personnel, we characterize current trends in T&E and provide several general estimating relationships that can be used early in program development to estimate T&E costs. Appendixes A and B briefly summarize relevant technical, schedule, and programmatic information on recent test programs, while Appendix C provides official definitions of the phases of T&E. A separate supplement provides corresponding cost information but is available only to authorized government personnel.

This project is being conducted within the RAND Project AIR FORCE Resource Management Program. The research is sponsored by the Principal Deputy, Office of the Assistant Secretary of the Air Force (Acquisition), and by the Office of the Technical Director, Air Force Cost Analysis Agency.

This monograph should be of interest to government cost analysts, the military aircraft and missile acquisition and T&E communities, and those concerned with current and future acquisition policies.

Other RAND Project AIR FORCE reports that address military aircraft cost estimating issues include the following:

- *An Overview of Acquisition Reform Cost Savings Estimates* (Mark Lorell and John C. Graser, MR-1329-AF) used relevant literature and interviews to determine whether estimates of the efficacy of acquisition reform measures are robust enough to be of predictive value.
- *Military Airframe Acquisition Costs: The Effects of Lean Manufacturing* (Cynthia Cook and John C. Graser, MR-1325-AF) examined the package of new tools and techniques known as "lean production" to determine whether it would enable aircraft manufacturers to produce new weapon systems at costs below those predicted by historical cost-estimating models.
- *Military Airframe Costs: The Effects of Advanced Materials and Manufacturing Processes* (Obaid Younossi, Michael Kennedy, and John C. Graser, MR-1370-AF) examined cost estimating methodologies and focus on military airframe materials and manufacturing processes. This report provides cost estimators with factors useful in adjusting and creating estimates based on parametric cost estimating methods.
- *Military Jet Engine Acquisition: Technology Basics and Cost-Estimating Methodology* (Obaid Younossi, Mark V. Arena, Richard M. Moore, Mark Lorell, Joanna Mason, and John C. Graser, MR-1596-AF) contains background information on modern aircraft propulsion technologies and a variety of cost-estimating methodologies.

RAND Project AIR FORCE

RAND Project AIR FORCE (PAF), a division of the RAND Corporation, is the U.S. Air Force's federally funded research and development center for studies and analyses. PAF provides the Air

Force with independent analyses of policy alternatives affecting the development, employment, combat readiness, and support of current and future aerospace forces. Research is performed in four programs: Aerospace Force Development; Manpower, Personnel, and Training; Resource Management; and Strategy and Doctrine.

Additional information about PAF is available on our Web site at http://www.rand.org/paf.

Contents

Figures

Tables

Summary

T&E is a key step in the development of any military weapon system. It is the primary means of ensuring that the system will actually perform its intended functions in its intended environment.

T&E of a modern weapon system is an involved and often lengthy process that reflects both the complexity of the system under test and the variety of specialized resources and activities its testing requires. T&E consumes a significant portion of the development time and resources for military aircraft and air-launched weapons,[1] which is why the general reexamination of acquisition processes that has taken place over the past decade has included T&E. Looking for efficiencies and cost savings, advocates of acquisition streamlining have questioned the scope, duration, cost, and organizational responsibilities of the traditional T&E process. These questions are even more urgent because most T&E expenditures occur in the later stages of development, when cost overruns and schedule slips from other activities may have become more apparent. As a result, there is often considerable pressure to expedite and/or reduce T&E activities to recoup some of the other overruns.

The T&E process has evolved with the complexity and cost of the systems being developed and with the priorities and practices of defense acquisition management. This evolution and its effects on the development cost of the systems under test are the subject of this monograph.

---

[1] On average, contractor and government T&E account for approximately 21 percent of development costs for fixed-wing aircraft and 15 percent for guided weapons.

The tasking for this study arose from two concerns. Some program managers have proposed test programs of greatly reduced scope and duration, citing such initiatives as increased use of modeling and simulation to reduce the amount of expensive "open air" testing. Other rationales for reduced test schedules and budgets include using lower-risk designs, combining government and contractor testing, using nondevelopmental item (NDI) and commercial-off-the-shelf (COTS) approaches, and applying total system performance responsibility (TSPR) contracting. Acquisition decisionmakers needed to know whether these approaches can achieve the projected savings.

The second concern was that members of the cost analysis community, particularly those outside of the program offices and test organizations, were not confident that the data and relationships they were using to estimate the costs of testing for a program or to cross check such estimates reflected the current T&E environment. Since some of their tools were based on development programs that were 15 to 30 years old, validation against current and evolving T&E approaches became a priority.

Although the original intention was for this study to focus on fixed-wing aircraft, the Air Force Cost Analysis Agency (AFCAA) asked RAND Corporation to include a cross section of tactical missiles and guided munitions. Because many of the programs of interest were joint Air Force–Navy development efforts and because the Navy cost community had similar requirements, the Assistant Secretary of the Navy for Research, Development, and Acquisition (ASN RDA) agreed and directed the appropriate Navy program executive officers and test activities to support the project.

The project scope involved the following four tasks:

- analyzing the nature of current T&E costs for aircraft, tactical missile, and guided munition systems and the trends likely to affect these costs in the immediate future
- identifying key cost drivers
- collecting, normalizing, and documenting representative data
- developing a set of practical, documented methodologies for making high-level T&E estimates.

To interpret the results of this study correctly, certain limitations and constraints should be kept in mind. First, the study focused on system-level testing associated with development programs funded through research, development, test, and evaluation and categorized as "system T&E." This therefore excluded postproduction follow-on testing, production acceptance testing, and component-level testing.

Second, the study focused only on what government program offices typically pay for, the items test organizations often refer to as *reimbursable costs*. These could be considered the price the customer pays for test services. These T&E costs are part of each weapon system's development budget, whether it is the contractor or the program office that directly incurs them.[2]

Third, we limited our analysis to recent Air Force and Navy fixed-wing aircraft, tactical missile, and guided munition programs. Because the purpose of the study was to examine current test practices, we focused generally on programs that had completed development within the past ten years or, in a few cases, slightly earlier.[3] Older data were used for trend analysis and, where appropriate, to augment more-recent data in developing relationships.

Because fewer new development programs are projected for the future, we attempted to include several programs representing major modifications to existing systems for which enough data were available for our analysis. Relevance to both modification and new development programs was also a consideration in selecting parameters for cost relationships.

Since our purpose was to examine the cost of testing as it was being conducted at the time of our research, we did not assess the efficiency and effectiveness of test procedures.

---

[2] The government does pay other T&E expenses, such as overhead and construction at test facilities, through specific appropriations. These are not allocated to any weapon system and, therefore, are not included in this study.

[3] One of the purposes of the study was to provide more-current cost, technical and programmatic data to the cost community. In a few cases, the data we collected were slightly older than our nominal 10 years but were not always generally available within the cost organizations and thus would be a useful resource.

Cost-Estimating Data, Methodologies, and Trends

To develop cost-estimating methodologies, we collected actual T&E costs, schedules, and programmatic test data from a number of sources, including the contractor cost data reports, system program offices, government cost analysis agencies, government test organizations, and selected contractors (see Acknowledgments). Chapter Four discusses these data, which we treat more fully in a limited-distribution supplement. The appendixes include detailed programmatic data on 16 aircraft and guided-weapon programs for reference.

Chapter Five presents the T&E cost estimating relationships (CERs) we developed from these data. The CERs and the data in the proprietary supplement should allow an estimator to compare estimates for a proposed program with actuals from other programs. Of course, the estimator will have to use expert judgment to take into account any specific, unique aspects of the proposed program. Chapter Five includes CERs for

- overall contractor test costs for aircraft
- contractor ground, flight, and "other" test costs for aircraft
- total contractor and government test costs for guided weapons.

As with most parametric estimating tools, these would be most useful for a Milestone B or earlier cost or test estimate, when fewer details of a proposed program are known. As the system progresses through development and more information becomes available, more-detailed estimating techniques can be used, with these CERs providing a cross-check at an aggregate level.

It was much more difficult to collect and document data on the costs government organizations had incurred than on corresponding contractor costs. We initially did not consider this to be a serious limitation, assuming that, because of acquisition reform, government costs would decrease as contractors took on a greater share of the effort. However, in cases where we were able to obtain government costs for programs, this generally did not prove true. Government T&E costs were substantial and, for guided weapons, generally greater than those of the system contractor. In many cases, contractors still rely on government test facilities and functional expertise, particularly for high-cost, low-utilization test capabilities. Government personnel normally participate actively in the integrated test teams. Even when the contractor can select any test facility it considers appropriate, that might end up being a government facility—with the government facility then becoming test subcontractor to the system prime contractor. Of course, most open-air testing continues to be conducted on DoD ranges.

Consistent accumulation and reporting of government cost data, to standards similar to those for contractor data, would greatly improve the accuracy of cost estimates for testing. This would ensure that the total program financial picture was available for management in the present and for analysis in the future. This would improve the ability of government test facilities to evaluate the cost and schedule implications of their processes, assess the contributions of all their activities, and focus investment and management attention on the activities most critical to each facility's customer base.

Overall T&E Cost Trends

The overall cost of T&E to the program shows no clear trend upward or downward over the last 20 to 30 years. Although government and industry test personnel have indicated that the increasing use of modeling and simulation, improvements in instrumentation and test processes, reduction of redundant testing, and various acquisition streamlining initiatives have reduced the cost of individual tests, other changes appear to have offset any potential net savings.

Thus, the proportion of development costs dedicated to T&E has remained relatively constant for aircraft and guided weapon systems. Although various explanations for this are possible, the dominant factors are probably the increasing complexity of the systems tested and the increasing content of test programs. (See the Cost Trends section in Chapter Three.)

T&E Issues and Findings

Another principal objective of this study was to identify changes in the practice of T&E and, to the extent possible, their likely effects on the cost of T&E for future aircraft, missiles, and guided munitions.

Overall, we found no cost or schedule data that would allow us to quantify how these practices individually affect current systems, either as upward or downward influences on test costs or schedules. The following paragraphs outline the issues we addressed.

Acquisition Reform

Acquisition reform initiatives are a diverse array of ideas, processes, and practices designed to streamline the DoD acquisition process, reducing either cost or schedule, or improving technology. A previous RAND report (Lorell, 2001) addressed the general effects of acquisition reform on cost estimating.

One of the acquisition reform initiatives that report discusses is TSPR, which transfers certain T&E responsibilities from DoD to the contractors. Although the data to support cost savings tend to be anecdotal, it is apparent that it will shift costs from government activities to contractor activities and must be recognized as such in future cost estimates. Our interviews suggest that TSPR must be well planned to avoid two test-related problems: Test data may not be available to DoD for other, postdevelopment uses, and cross-platform integration might not be adequately coordinated, especially in guided weapon development. DoD must have the foresight to ensure that it can use the system development and demonstration test data to design modifications or to qualify additional platform-and-weapon configurations. In addition, to maintain compatibility, DoD will have to ensure careful coordination of cross-platform integration issues, particularly with guided-weapon development and modification, with other systems.

It is too early to assess the outcome of recent innovative test management approaches that give the contractor broad latitude in developing and executing the developmental test program. Another innovative approach, relying on non-DoD tests and certifications of nondevelopmental aircraft for DoD applications, was not generally as successful as its promoters had hoped. We found that Federal Aviation Administration certification alone is not sufficient to demonstrate that a particular aircraft meets most military performance specifications. "Best commercial practices" are not an effectively codified set of procedures, like common law or accounting principles. Because they tend to be situational and inconsistent from contractor to contractor, they may be inadequate for responsible acceptance of military systems. (See the Acquisition Reform section in Chapter Three.)

Modeling and Simulation

Virtually all test programs now incorporate modeling and simulation. In many programs, some aspects of the analytical tools have not been mature enough to give enough confidence for waiving live testing. However, in all cases, modeling and simulation at least reduced the risk, and often the duration, of live tests and thus appeared to be a good investment. In addition to directly benefiting T&E, robust modeling and simulation also benefits

- evaluating design excursions during development
- tactics development
- operator training
- evaluating future system enhancements. (See the Modeling and Simulation section in Chapter Three.)

Testing of Software-Intensive Systems

An area of almost universal concern was effective testing of software intensive systems, which are growing in complexity and functionality. Continuing advances in technology have translated into system capabilities unimagined a generation ago. The growth in capability translates into increased test complexity. This area should receive specific attention in any future T&E estimates. (See the Software Intensive Systems section in Chapter Three.)

Combined and Multiservice Testing

There was general agreement that integrated contractor-government test teams were a positive force in optimizing testing. Similarly, combined development and operational test teams have been valuable because they avoid redundant testing and highlight operational effectiveness and suitability issues for early resolution. Some program personnel expressed a desire for even more intensive "early involvement" by the operational test community. The primary constraint appears to be limited staffing of the service operational test organizations. (See the Combined and Multiservice Testing section in Chapter Three.)

Contractor Versus Government Test Facilities

While there was general agreement that the major government test facilities are essential for executing the required test programs and that they generally provide excellent support, some contractor personnel expressed varying levels of frustration in their dealings with the government test organizations. In programs with aggressive affordability goals, there was a concern that some government test range personnel were not as focused on controlling the costs and schedule of the test program as other members of the test team were. Some felt that there were practices at the ranges that were overly conservative and caused unnecessary costs and delays. In other cases, delays resulted from chronic understaffing or procedures with little provision for flexibility. These issues are of increasing importance when contractors are given incentives to perform within what are, in effect, fixed test budgets and schedules. A related contractor concern was that the government ranges tended to be "overfacilitized" but "undermodernized." (See the Contractor Versus Government Test Facilities section in Chapter Three.)

Live-Fire Testing

Although live-fire testing can be a contentious issue during early planning for system development, our interviews did not highlight major concerns at the program level, as long as the requirements were known in advance and planned for accordingly. Because data were limited, we could draw no general conclusions about the real cost of live-fire testing. (See the Live-Fire Testing section in Chapter Three.)

Although there is some disagreement over the appropriate level of testing in specific circumstances—live-fire testing, testing for statistically rare events, etc.—we found little controversy in general over the scope of testing. Other studies have concluded that most DoD test programs have already eliminated the majority of unnecessary or redundant testing. Several sources, however, expressed the opinion that thoughtful reevaluation of test procedures could improve the pace and efficiency of the typical test program. (See the Live-Fire Testing and the Contractor Versus Government Facilities sections in Chapter Three.)

Warranties

None of our interviews indicated that warranties significantly changed the T&E process or costs.

Acknowledgments

In performing this study, the authors had extensive discussions with subject-matter experts in both government and industry. Indeed, without their willingness to share their time, insights, and data generously, this study would have been more limited in both scope and utility. To all who so willingly supported this effort, we offer our sincere thanks. Unfortunately, they are far too numerous to mention individually, but we would like to specifically acknowledge the principal organizations visited and our key points of contact in each:

- Air Force Cost Analysis Agency: Joseph Kammerer, Director; Jay Jordan, Technical Director
- Director of Test & Evaluation, Headquarters, USAF (AF/TE): John T. Manclark
- Office of the Secretary of Defense (OSD/PA&E): Gary Bliss
- Director, Operational T&E: Christine Crabill
- 46th Test Wing: Carlos Rodgers
- Aeronautical Systems Center: Michael Seibel
- Air Armament Center: Jung Leong
- Air Force Flight Test Center: James Dodson; Vicky Yoshida
- Air Force Operational Test Center: Joseph Guy
- Chief of Naval Operations (N091): Skip Buchanan
- Commander, Operational T&E Force: Commander William Padgett
- Naval Air Systems Command: David Heller (AIR-5.0); David Volpe (AIR-4.2)

- Naval Air Warfare Center—Aircraft Division: J. R. Smullen; Robert Mann
- Naval Air Warfare Center—Weapons Division: Robert Ostrom; Robert Copeland
- Naval Center for Cost Analysis: Bill Stranges
- B-1B: Rick Wysong
- B-2: Larry Perlee
- C-17: Eric Wilson
- F-16: Douglas Egged
- F-18: Steve Kapinos
- F-22: Melanie Marshall; Mark Whetstone
- JPATS: Jay Free
- Joint Strike Fighter: Paul Tetrault
- T-45: Lieutenant Commander Jennifer Rigdon
- V-22: Sandie Raley
- AIM-9X: Toby Jones
- AMRAAM: Robert Guidry; Jack Trossbach
- JASSM: Commander Patrick Roesch
- JDAM: Mike Evans
- JSOW: Michael Chartier
- Sensor Fuzed Weapon: Beth Crimmins
- SLAM-ER: Carl Smith
- Tomahawk: Lieutenant Commander Tim Morey
- WCMD: Duane Strickland
- Boeing Military Aircraft and Missile Systems: Kurt Syberg
- Lockheed Martin Aeronautics: Paul Metz
- Raytheon Missile Systems: Glen Pierson.

We would also like to thank several individuals in particular whose support was both extensive and central to the completion of this study: Jay Jordan for his guidance and helpful suggestions; Dave Heller, whose encouragement and persistence opened many Navy doors; Bill Stranges, who provided both thoughtful feedback and missing data; and Steve Cricchi, Joe Guy, Bob Kellock, CDR Bill Padgett, Larry Perlee, Kurt Syberg, Paul Tetrault, Jack Trossbach, Joe

Wascavage, Eric Wilson, and Vicky Yoshida, whose insights and efforts in locating key data were indispensable.

Our RAND colleagues Michael Kennedy and Dave Stem improved the report significantly by their thorough review and constructive comments. Tom Sullivan assisted with the data analysis. Michele Anandappa provided invaluable administrative support.

Abbreviations

AAC	Air Armament Center
AAV	AMRAAM Air Vehicle
AAVI	AMRAAM Air Vehicle Instrumented
ACAT	acquisition category
ACE	AMRAAM captive equipment
ACTD	advanced concept technology demonstration
AEDC	Arnold Engineering Development Center
AFB	air force base
AFCAA	Air Force Cost Analysis Agency
AFMSS	Air Force Mission Support System
AFOTEC	Air Force Operational Test and Evaluation Center
AIAA	American Institute of Aeronautics and Astronautics
AMRAAM	Advanced Medium-Range Air-to-Air Missile
AoA	analysis of alternatives
APREP	AMRAAM Producibility Enhancement Program
ASN (RDA)	Assistant Secretary of the Navy for Research, Development, and Acquisition
BAe	British Aerospace

CAIV	cost as an independent variable
CARD	cost analysis requirements description
CBU	cluster bomb unit
CCDR	contractor cost data report
CCRP	captive-carry reliability program
CDR	critical design review
CER	cost estimating relationship
CLV	captive-load vehicle
CMUP	Conventional Mission Upgrade Program
COI	critical operational issue
COMOPTEVFOR	Commander, Operational Test and Evaluation Force
COTS	commercial-off-the-shelf
CPR	cost performance report
CTD	concept and technology development
CTF	combined test force
CTU	captive-test unit
DEM/VAL	demonstration and validation
DoD	Department of Defense
DOT&E	Director of Operational Test and Evaluation
DT	developmental testing
DT&E	development test and evaluation
EAC	estimate at completion
EDM	engineering development missile
EMC	electromagnetic compatibility
EMD	engineering and manufacturing development
EMI	electromagnetic interference
EW	electronic warfare

FAA	Federal Aviation Administration
FDE	force development evaluation
FOT&E	follow-on operational test and evaluation
FMS	foreign military sales
FSD	full-scale development
FTTC	Flight Test Technical Committee
FY	fiscal year
GAM	GPS-aided munition
GPS	Global Positioning System
GTV	guided test vehicle
HITL	hardware in the loop
ILS	instrument landing system
IMV	instrumented measurement vehicle
INS	inertial navigation system
IOC	initial operational capability
IOT&E	initial operational test and evaluation
IPT	integrated product team
IR	infrared
IRT	instrumented round with telemetry
ISTF	installed system test facility
IT	integrated testing
JASSM	Joint Air-to-Surface Standoff Missile
JDAM	Joint Direct Attack Munition
JET	joint estimating team
JHMCS	Joint Helmet Mounted Cueing System
JPATS	Joint Primary Aircraft Training System
JSF	Joint Strike Fighter
JSOW	Joint Standoff Weapon
LFT&E	live-fire test and evaluation

LO	low observability
LRIP	low-rate initial production
M&S	modeling and simulation
Mk.	mark
MNS	mission needs statement
MOE	measure of effectiveness
MOT&E	multiservice operational test and evaluation
MRTFB	Major Range and Test Facility Base
NAVAIR	Naval Air Systems Command
NAWC-AD	Naval Air Warfare Center–Aircraft Division
NAWC-PM	Naval Air Warfare Center–Pt. Mugu
NDI	nondevelopmental item
OAR	open-air range
OPEVAL	operational evaluation
OPTEVFOR	Operational Test and Evaluation Force
ORD	operational requirements document
OSD	Office of the Secretary of Defense
OT	operational testing
OTA	Operational Test Agency
OT&E	operational test and evaluation
P3I	preplanned product improvement
PD/RR	program definition/risk reduction
PEP	Producibility Enhancement Program
PID	program introduction document
PRTV	production representative test vehicle
QOT&E	qualification operational test and evaluation
QT&E	qualification test and evaluation
RCS	radar cross section
RDT&E	research, development, test, and evaluation

SAF/AQ	Assistant Secretary of the Air Force (Acquisition)
SCTV	separation control test vehicle
SDD	system development and demonstration
SFW	sensor fuzed weapon
SLAM-ER	Standoff Land-Attack Missile–Expanded Response
SOC	statement of capability
SPO	system program office
STARS	Standard Accounting and Reporting System
ST&E	system test and evaluation
STU	seeker test unit
STV	separation test vehicle
T&E	test and evaluation
T1	theoretical first unit
TAMPS	Tactical Automated Mission Planning System
TASSM	Triservice Air-to-Surface Attack Missile
TECHEVAL	technical evaluation
TEMP	test and evaluation master plan
TSPR	total system performance responsibility
UJFT	Undergraduate Jet Flight Training System
USC	United States Code
VCD	verification of correction of deficiencies
VOR	very-high-frequency omnidirectional range (station)
WBS	work breakdown structure
WCMD	Wind-Corrected Munitions Dispenser
WIPT	working-level integrated product team

CHAPTER ONE

Background: The Cost of Testing

Test and evaluation (T&E) is a key step in the development of any military system. It is the primary means of ensuring that the system will actually perform its intended functions in its intended environment.

T&E of a modern weapon system is an involved and often lengthy process reflecting both the complexity of the system under test and the variety of specialized resources and activities required. As Figures 1.1 through 1.4 show, for military aircraft and air-launched weapons, T&E represents a significant portion of development resources and time.[1] Because of this, T&E has been included in the general reexamination and attempts to improve acquisition processes conducted over the past decade. Looking for efficiencies and cost savings, advocates of acquisition streamlining have questioned the scope, duration, cost, and organization of the traditional T&E process. These questions are even more urgent because most T&E expenditures occur in the later stages of development, when cost overruns and schedule slips from other activities may have already occurred. As a result, although T&E may or may not be the pacing activity in a program's execution, there is often considerable pressure to reduce T&E activities to compensate for losses due to overruns and slips.

---

[1] To keep Figures 1.1 and 1.2 nonproprietary but to give the reader a sense of the magnitude of the test costs, we divided total contractor and government test costs by the total full-scale development (FSD) and engineering and manufacturing development (EMD) costs for each system.

Figure 1.1
Aircraft System T&E Share of Total FSD and EMD Costs

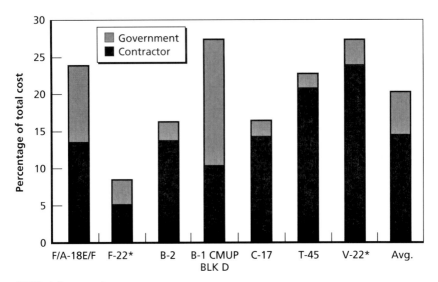

NOTE: * Program in progress
RAND *MG109-1.1*

Figure 1.2
Guided-Weapon System T&E Share of Total FSD and EMD Cost

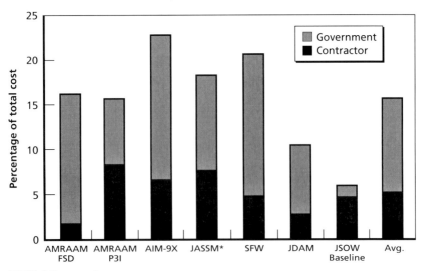

NOTE: * Program in progress
RAND *MG109-1.2*

Figure 1.3
DT and OT Share of Total Aircraft FSD and EMD Duration

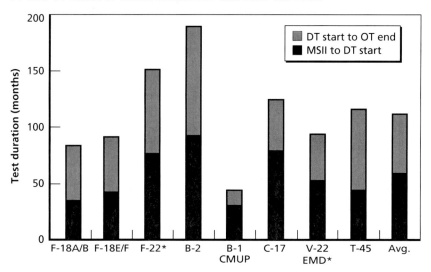

NOTE: * Program in progress
RAND *MG109-1.3*

Figure 1.4
DT and OT Share of Total Guided-Weapon FSD and EMD Duration

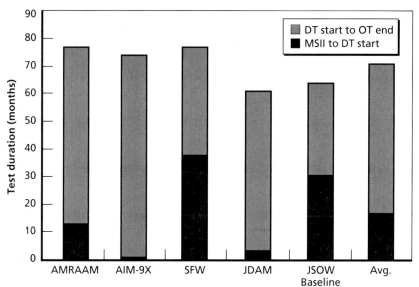

RAND *MG109-1.4*

The T&E process has evolved with the complexity and cost of the systems being developed and with the priorities and practices of defense acquisition management. This evolution of the T&E process and its effect on the development cost of the systems under test are the subject of this monograph.

The tasking for this study arose from two concerns. The first was the desire of senior acquisition managers to get a satisfactory answer to a seemingly straightforward question: "Are we budgeting an appropriate amount for T&E in this program?" With the pervasive emphasis on acquisition streamlining, commercial practices, cost-as-an-independent-variable (CAIV) approaches, and a variety of lean initiatives, there was a concern that the current T&E process might be a holdover from when contractors built hardware to detailed government specifications and standards and when redundant, stove-piped processes increased both cost and schedule.

On the other hand, program managers, under intense affordability pressures, were reexamining every aspect of their programs and looking for potential cost savings. As a result, some program managers were proposing to reduce the scope and duration of test programs greatly, citing such initiatives as increased use of modeling and simulation (M&S) to reduce the amount of expensive "open air" testing. Other rationales for reduced test schedules and budgets include using lower-risk designs, combining government and contractor testing, using nondevelopmental item (NDI) and commercial-off-the-shelf (COTS) approaches, and applying total system performance responsibility (TSPR) contracting.

The second concern was that members of the cost analysis community, particularly those outside of the program offices and test organizations, were not confident that the data and relationships they were using to provide early estimates of the costs of testing for a program or to cross check such estimates reflected current trends in the current T&E environment. Since some of their tools were based on development programs that were 15 to 30 years old, validation against current and evolving T&E approaches was a priority.

Changes in the Test Process

Independent of specific budget and schedule pressures, the approach to T&E should be consistent with the streamlined approaches that have been introduced into other areas of defense acquisition management.

A major thrust of acquisition reform has been to integrate and streamline what were perceived as stovepiped processes, to reduce the time and cost they involved. In particular, such initiatives as integrated product teams (IPTs) were established to make planning and execution of tests more efficient, to improve communications among the various disciplines and organizations involved, and to improve their ability to respond to unplanned events. In T&E, this begins with combining contractor test personnel and their government counterparts into an integrated developmental testing (DT) team. The intent is to ensure that the contractor, who generally has lead responsibility for executing the DT program, meets the government DT data requirements with sufficient visibility during tests to allow the program to progress. This parallels the IPT organization in other areas of defense acquisition.

A second integration initiative was the formation of combined test forces (CTFs). These allow the operational testing (OT) community to participate substantially in DT activities that formerly focused on proving engineering designs rather than operational effectiveness. This has two desirable results:

- Involving the OT community helps identify potential operational effectiveness and suitability issues early, when corrective measures are generally less disruptive and less expensive.
- Maximizing the ability of the OT community to use data developed during DT has the potential to reduce or eliminate redundant dedicated OT.

Another trend in defense acquisition that has affected T&E has been the use of a reduced number of top-level performance specifications rather than detailed item specifications. This requires the test community to design test programs that both demonstrate the achievement of critical performance parameters and minimize the time and resources needed to do so.

Trends in technology have had the most obvious effects on the way T&E is conducted today. The systems under test are considerably more complex than those even one generation older. This has required a corresponding improvement in the capabilities of testing tools and methods. The increasing use of M&S is an obvious example of how technology has improved testing capability and productivity. What may not be so obvious is that these tools are often necessary because live testing of all functions of current systems is impractical, if not impossible.

T&E Constituencies

Different constituencies bring diverse perspectives to T&E (see Table 1.1). Depending on the program phase and how the data are to be used, T&E can be either a learning process or the final exam. Optimizing T&E for each constituency would tend to drive T&E priorities and objectives in different directions. For example, system designers tend to view T&E as an integral part of the development process. Depending on the type of test, designers may view it as an experimental confirmation of the engineering design approaches or a realistic exercise of a complex system to discover incompatibilities or integration problems. In most cases, designers and engineers approach T&E as a learning process; the more rigorous the test, the more you learn, but the more likely you are to fail. This also tends to be the approach in commercial product testing because "in-service" product failures can adversely affect future sales.

Table 1.1
T&E Perspectives

Role	T&E Objective
Designer	Insight into expected versus actual performance
Manager	Evidence of progress, design maturity, risk reduction
User	Assurance that the system can perform its intended mission

Managers, who are concerned with schedules and budgets, tend to view T&E as a series of milestones that signify graduation from one phase of development to another. Rigorous T&E consumes resources and time and must therefore be carefully structured, managed, and optimized so standards are met without overtesting. From the management perspective, failure can lead to costly redesign and retesting. One comment that the Director of Operational Test and Evaluation (DOT&E) has heard is that a "dollar spent on T&E is a dollar spent looking for trouble" (Jonson, 2002). Yet the Defense Science Board Task Force on T&E found that "[i]n most DoD [Department of Defense] test programs, the content is already at or near minimum" (Defense Science Board, 1999).

The ultimate users of the system tend to view T&E (especially OT) as their assurance that the system will actually perform its mission in the "real world." Theirs is the concern that a system that performs its functions in a limited, possibly idealized environment may be purchased and fielded yet fail when subjected to operation and maintenance under field conditions. While designers and managers tend to focus on compliance with technical specifications, operational testers are primarily concerned with end-to-end testing against the ultimate operational requirements.

Estimating the Cost of T&E

The cost of a T&E program can be estimated in several ways. The specific approach taken is normally a function of the time and data available to do the cost estimate, as well as the organization directing the estimate.

The initial T&E estimate is generally developed as part of the planning budget for the development program. As in many estimating situations, initial estimates are often based on comparisons with the actual cost of similar predecessor programs. This approach depends on the timely availability of cost and characteristic data of the analogous programs. If the programs are similar overall, the cost data may be adjusted in those areas where the similarities are not as strong to better represent the characteristics, and presumably the cost, of the program being estimated.

A second approach to estimating the cost of T&E for a new system is parametric analysis. Analysts collect cost data from multiple programs and compare them through statistical tests to uncover cost-driving program characteristics. This approach requires not only that historical cost, technical, and programmatic data be available for multiple programs but also that the data set either be homogeneous enough or have enough data points to permit statistical stratification by class. The simplest parametric relationships represent T&E cost as a constant multiple of a single cost-driving parameter (for example, development cost, number of flight-test hours). Given sufficient data, analysts can develop more-robust cost-estimating relationships (CERs) statistically using multiple parameters.

The third approach is to sum lower-level costs estimated by various methods. This method is typically used after test plans have become available and once the types and quantities of test resources have been identified. Primary test organizations typically do this type of estimating, working from the test requirements system program office (SPO) has provided. Normally the estimator uses cost rates per test resource unit (for example, number of flight-test hours, type of data collection, target or threat presentations) to estimate the cost for the testing required. Test organizations typically set their rates annually. This method requires accurate descriptions of the program requirements and estimators with the experience to make allowances for the inevitable scrubs, delays, and retests. This detailed, build-up approach has the advantages of accounting for the unique requirements of a particular program and of being useful for setting and tracking budgets. Its disadvantage is that it is based on assessments of future test requirements, which may not account for contingencies or may otherwise underscope effort or content.

Study Objectives

The Office of the Assistant Secretary of the Air Force (Acquisition) (SAF/AQ) tasked RAND Project AIR FORCE to analyze the major categories of acquisition support costs (commonly referred to as "below-the-line costs") to improve the Air Force cost-analysis community's estimating approaches and tools. Discussions with the project technical monitor at the Air Force Cost Analysis Agency (AFCAA) indicated that the most pressing need was to examine system T&E costs. Although the original intention was to focus on fixed-wing aircraft, AFCAA asked RAND to include a cross section of tactical missiles and guided munitions as well. Since many of these programs were joint Air Force–Navy development efforts and since the Navy cost community had similar requirements, SAF/AQ requested Navy support and participation through the Office of the Assistant Secretary of the Navy for Research, Development, and Acquisition (ASN [RDA]). ASN (RDA) agreed and directed the appropriate Navy program executive officers and test activities to support the project.

The project involved the following four tasks:

- analyzing the nature of current aircraft, tactical missile and guided munition system T&E costs and trends likely to affect them in the immediate future
- identifying key cost drivers
- collecting, normalizing, and documenting representative data
- developing a set of practical and documented cost-estimating methodologies.

Study Limitations

To interpret the results of this study correctly, certain limitations and constraints should be kept in mind. First, the study includes only those costs typically paid for by the program offices (which the test organizations often refer to as *reimbursable costs*). These costs could be considered the "price to the customer" for test services. Specifically, we did not collect or analyze cost data on the internal costs or operations of the DoD test infrastructure.[2] The program-funded test costs are the ones that are included in weapon system estimates prepared for service and Office of the Secretary of Defense (OSD) reviews and that are included in program budgets. Direct appropriations cover the costs of test organizations' facilities and infrastructure rather than the "variable" costs of testing. A variety of studies and panels have been examining the broader issues of test infrastructure.

We limited the programs we analyzed to recent Air Force–Navy fixed-wing aircraft, tactical missiles, and guided munitions. We focused on programs that had completed development within the past ten years or, in a few cases, slightly earlier, if test costs and programmatic data for those programs were not readily available to DoD cost analysts.[3] Older data from existing reports was used for trend analysis and, where appropriate, to augment more-recent data in developing relationships.

In an effort to consistently define testing to be included in the study, we also limited ourselves to the testing categorized as system T&E and to that associated with RDT&E-funded development programs. Thus, the study results do not capture subsystem testing (such as for a landing gear or an individual avionics component), which would normally be included in what the prime contractor pays the supplier for the subsystem. Cost-collection systems do not normally identify such costs separately. However, the study includes prototype testing if the activities were of sufficient scope and would have otherwise been included in the EMD phase (now called system development and demonstration [SDD]).[4] On the other hand, various types of postproduction follow-on testing and production acceptance testing were not included. We also did not collect data on

---

[2] These are costs often referred to as *direct budget authority*, since they are paid for using funds appropriated for T&E in the DoD budget.

[3] One of the purposes of the study was to provide more-current cost, technical, and programmatic data to the cost community. In a few cases, the data we collected were slightly older than our nominal 10 years but were not generally available within the cost organizations and thus would be a useful resource.

[4] T&E costs from the demonstration and validation (DEM/VAL) phases of the F-16 and AV-8B were included in the FSD/EMD totals because of their scope.

advanced concept technology demonstration (ACTD) programs, although we did discuss them in our interviews.

Because there are expected to be fewer new development programs in the future, we attempted to include several programs representing major modifications to existing systems, when enough data were available. This was also a consideration in selecting parameters for cost relationships.

Since the purpose of the study was to document the cost of testing as it is currently being conducted, test procedures or test sufficiency were not assessed.

Organization of the Report

Chapter Two provides a basic overview of T&E for cost analysts. The topics include:

- the T&E process
- types of T&E
- process flow and documentation
- resources and facilities.

Chapter Three summarizes trends affecting DoD testing and is based on our interviews with Air Force and Navy test and program personnel. Chapter Four describes how T&E costs are accounted for and what the typical data sources are and provides an analysis of cost trends in DT and OT. Chapter Five recommends approaches for estimating or assessing the realism of program system T&E costs, while Chapter Six offers conclusions and other recommendations. Appendixes A and B consist of brief summaries of the DT and OT of recent aircraft and guided-weapon programs, respectively. Appendix C offers an excerpt from the relevant military handbook.

TR-114-AF is a limited-distribution supplement to this report containing proprietary cost data for the programs described in Appendixes A and B.

CHAPTER TWO
The T&E Process

T&E involves the collection and analysis of data on the actual or projected performance of a system or its components. As Chapter One explained, T&E has three basic functions:

- providing designers with insight on the relationship between expected and actual performance
- helping managers assess design maturity
- assuring prospective users that the system can perform its intended mission.

The testing process can be divided into two major categories: developmental and operational. Developmental testing (DT) is performed at the part, subsystem, or full system level to prove the validity or reliability of the design, materials used, etc. The results of DT are used to modify the system design to ensure that it meets the design parameters and system specifications. Although operational testing (OT) relies in part on the results of DT, it is designed to test a system in its operational environment, where operational personnel (rather than technicians) would be responsible for operating, maintaining, and repairing the system in a realistic environment.

The intensity and duration of T&E activities vary as the program progresses through the acquisition process (see Figure 2.1). In the concept and technology development (CTD) phase, the T&E working-level IPT (WIPT) is formed, and an evaluation strategy is developed to describe the early T&E approach for evaluating various system concepts against mission requirements. M&S activities also begin at this time.

Figure 2.1
T&E Phases Within the Acquisition Process

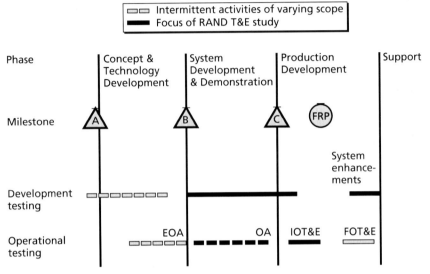

FRP = full-rate production.
EOA = early operational assessment.
OA = operational assessment.
RAND *MG109-2.1*

During the SDD phase, T&E focuses on evaluating alternatives; components; and eventually, system performance and suitability, as specified in the mission needs statement (MNS), the operational requirements document (ORD), the acquisition program baseline, and the T&E master plan (TEMP).

Once the system enters the production and deployment phase, full system-level testing predominates. Key developmental, operational, and live-fire tests of production-representative articles must be complete before a decision to move beyond low-rate initial production (LRIP) can be made. Follow-on OT is conducted to complete any deferred testing, verify correction of deficiencies uncovered in previous testing, and refine operational employment doctrine.

T&E continues into the operations and support phase to support evolutionary or block upgrades. A large program may have a standing T&E organization, located at a government test center, to ensure continuity, maximize test program efficiency, and smooth transitions from upgrade to upgrade.

Congress, in its role of funder of DoD programs, is among the organizations the T&E process must satisfy and has incorporated its requirements into law. For example, 10 U.S.C. 2399 specifies that organizations independent of the developing activity must conduct realistic OT before a program can proceed beyond low-rate production. DOT&E has oversight of this process and must submit annual reports to Congress.

Starting in 1987, Congress has also required that major systems and weapon programs undergo realistic survivability and lethality testing before proceeding beyond low rate production (10 U.S.C. 2366). Later sections in this chapter describe operational and live-fire testing in greater detail.

Defense acquisition, which includes T&E activities, has continued to evolve since the early 1990s as increasing pressures on defense budgets have driven the rationalization and streamlining of the process. The current emphasis is on time-phased system requirements so that improvements in capabilities are added incrementally, as supporting technologies mature. This strategy, often called *evolutionary acquisition* or *spiral development*, presumably reduces costs, risks, and development time.

Another goal of many recent acquisition reform initiatives has been increasing the flexibility of the process. While some statutory and regulatory requirements remain, emphasis on tailoring the process to fit individual programs has increased. As an integral part of the development process, T&E is also affected by these changes. The effects may range from eliminating duplicative testing or reporting to allowing the system contractor wide latitude in planning and executing DT, with the government specifying only the system-level performance the contractor must demonstrate.

The sections below and Chapter Three discuss specific T&E process initiatives. Since additional changes to the DoD acquisition process were being drafted as this monograph was being completed, the DoD acquisition directives should be consulted for the most current process information.

Types of Testing

Developmental Test and Evaluation

Developmental T&E (DT&E) is the T&E conducted by the developing organizations to determine whether the system meets its technical and performance specifications. The DoD program manager has overall responsibility for DT&E. With acquisition reform, the system contractor performs most detailed planning and test execution with the oversight of the T&E WIPT. The T&E WIPT normally includes representatives from

- the government program office
- the system contractor
- government test facilities
- the OT activity
- other organizations participating in or supporting testing.

The T&E WIPT is established during the CTD phase or, if there is no CTD phase, before the program is designated an acquisition program. One of the WIPT's first duties is to plan the approach to be used for evaluating system concepts against the requirements specified in the ORD or MNS and for the use of M&S. The resulting *evaluation strategy* also serves as a basis for the TEMP.

The TEMP is submitted before Milestones B and C for all acquisition category (ACAT) I programs.[1] This document describes the structure, objectives, activities, schedule, and resources required to execute the planned test program. It is also updated before OT periods.

---

[1] An ACAT I program is one that the Secretary of Defense has designated as a major defense acquisition program. Such a program normally involves an expenditure of more than $365 million (FY 2000$) for research, development, test, and evaluation (RDT&E) or more than $2.19 billion (FY 2000$) for procurement (10 USC 2430).

The contractor normally develops an *integrated test plan* to document and coordinate the detailed test planning for all DT&E activities.

T&E activities take place throughout system development. DT&E supports the system engineering process in evaluating the feasibility and performance of alternative concepts and identifying risk areas. It supports designers by evaluating component-, subsystem-, and system-level performance, often highlighting areas for improvement. DT&E also provides managers with objective assessments of the progress and maturity of the development program. For programs using a performance-based acquisition approach, DT&E is the primary means of visibility into contractor progress and system performance. Both the OT community and users find the data generated during DT useful for projecting the eventual operational utility of the system.

A key role of the T&E WIPT is to plan and coordinate the DT&E requirements of all organizations to maximize the utility of test data generated and minimize unnecessary conduct or repetition of tests. One example of this is the elimination of a dedicated period of government technical evaluation (TECHEVAL) for most systems. Government DT organizations now use data generated during DT to support their evaluations. Dedicated government DT is normally used only for areas of government concern that the contractor does not or cannot test adequately. Likewise, integrating OT personnel into a CTF allows what is referred to as combined DT and OT, in which the tests support both objectives. Thus, OT activities can provide meaningful feedback earlier in the development process, as well as reducing or eliminating duplicative testing during the dedicated operational evaluation (OPEVAL). In general, it is preferable to meet government oversight requirements through independent evaluation of jointly collected data.

Another function of DT&E is to demonstrate that the development phase is essentially complete and that the system is ready for dedicated OT. The program manager must certify this readiness.

Finally, successful production qualification testing of low-rate LRIP articles before the full-rate production decision demonstrates that a mature production process is in place. DT&E continues at a lower level throughout the life of the program to evaluate both correction of deficiencies identified in testing or operation and system enhancements.

Operational Test and Evaluation

In operational T&E, service test organizations conduct tests in operationally realistic conditions against threat or threat-representative forces to determine whether the system is operationally effective and suitable. *Operational effectiveness* evaluates the ability of the system to perform its assigned mission. *Operational suitability* evaluates the ability of operational personnel to operate and sustain the system in peacetime and wartime environments and is a function of such characteristics as reliability, maintainability, availability, supportability, transportability, compatibility, and adequacy of proposed training procedures.

The Air Force Operational Test and Evaluation Center (AFOTEC) and the Navy's Operational Test and Evaluation Force (OPTEVFOR) are normally involved in system testing from the early stages of development. Their role is to provide feedback to developers on operational issues and to provide periodic operational assessments based on M&S and DT. Early operational assessments, which are done before the SDD phase, use studies, M&S, lab trials, demonstrations, and prototypes to evaluate alternatives and the level of risk and estimate military utility. Operational assessments, done during SDD, use engineering development models or production-representative systems. Critical operational issues (COIs) are the questions that OT must answer. These are derived from ORD requirements and give focus to OT planning and testing. Measures of effectiveness (MOEs) are used to evaluate performance on COIs.

Since most programs use what is termed *combined* DT and OT, planning for DT also considers OT objectives. The resulting tests can thus provide data useful for both DT and OT purposes. This minimizes expensive testing, while providing timely data for analysis of technical performance (DT) and operational performance (OT).

Before any ACAT I or II program can proceed beyond LRIP, it must meet a statutory requirement for independent testing of production or production-representative articles in an operationally realistic environment to demonstrate the system's performance for decision-makers.[2] The Air Force refers to this testing as *initial operational T&E* (IOT&E), and the Navy refers to it as *OPEVAL*. The appropriate OT agency conducts the tests in operationally realistic scenarios involving military personnel trained as users and maintainers. Statute prohibits the system contractor from participating in IOT&E in any roles other than those planned for the combat operation of the system.

Because OT personnel are extensively involved in earlier testing, dedicated IOT&E is normally much shorter than DT. In effect, IOT&E can be considered a "final exam" and is therefore a significant milestone in the system development. Because of its importance, the program manager is required to certify the system's readiness for dedicated OT.

While OT, by its nature, attempts to maximize realism, full-scale live testing is not practical in some circumstances. In these cases, the OT agency, with the concurrence of DOT&E, can use data from other testing, such as hardware-in-the-loop testing and M&S,[3] to independently evaluate likely operational effectiveness and suitability.

The OT agency's involvement does not stop with the completion of dedicated OT. It must forward an IOT&E report to the Secretary of Defense and Congress. The agency also must conduct follow-on operational T&E (FOT&E) after the full-rate production decision to

- finish any incomplete IOT&E testing
- verify correction of IOT&E deficiencies

---

[2] An ACAT II program is one that a service secretary estimates will require an expenditure of $140 million (FY 2000$) in RDT&E funds or more than $660 million (FY 2000$) in procurement but less than the thresholds for an ACAT I program (see footnote 3) (DoD 5000 Interim Guidance [since superseded by new guidance]).

[3] Hardware-in-the-loop testing involves exercising actual systems or subsystems in a controlled environment using simulated inputs.

- refine estimates or projections from IOT&E
- evaluate significant changes to the system design or employment
- evaluate new operational concepts or environment.

Additional OT is conducted whenever system modifications materially change that system's performance (DoD 5000.2-R). FOT&E usually involves the using command. DOT&E or the OT agency determines the quantity of test articles for dedicated OT.

Multiservice Operational Test and Evaluation

Systems that multiple services will acquire or use must undergo multiservice operational T&E (MOT&E). The designated lead service has primary responsibility for the test program and test procedures, with participation from the other services. A service with unique requirements does its own planning, testing, and funding for them. Because of differences in employment, test results that may be satisfactory for one service may not be for another.

Qualification Operational Test and Evaluation

Qualification operational T&E (QOT&E) is an Air Force term for the testing of modifications or new employment of a system for which there is no RDT&E-funded development.

Live-Fire Test and Evaluation

Although live-fire testing of manned systems for survivability (a combination of susceptibility and vulnerability) and missile and munitions for lethality had previously been a part of system testing, the 1987 Defense Authorization Act mandated that all "covered systems" undergo realistic full-scale live-fire testing before proceeding beyond LRIP. DoD 5000.2-R defines *covered system* as

- a major system within the meaning of that term in 10 U.S.C. 2302(5) that is,
  - user-occupied and designed to provide some degree of protection to its occupants, or
  - a conventional munitions program or missile program; or

- a conventional munitions program for which more than 1,000,000 rounds are planned to be acquired; or
- a modification to a covered system that is likely to affect significantly the survivability or lethality of such a system.

Under additional legislation in 1994, Congress directed DOT&E oversight of live-fire T&E (LFT&E). DOT&E may grant waivers when realistic system-level LFT&E would be "unreasonably expensive and impractical." DOT&E must approve waivers of full-scale live-fire testing in combat configuration before Milestone B or program initiation. The waiver request must describe the proposed alternative LFT&E strategy.

LFT&E involves testing at the component, subsystem, and system levels. It is normally part of DT to allow identification and implementation of any necessary design changes before full-rate production. Actual live-fire tests or shots are normally preceded by extensive modeling of damage or lethality to determine shot lines and sequence of shots to maximize the value of data collected and to assess model validity.

Test and Evaluation Process Steps

All effective T&E is tailored to the system performance requirements, the nature and maturity of the system under test, and the time and resources available. However, the complexity of testing a major weapon system and the coordination required dictate using a general process or framework as a guide for effective planning and execution of the test program.

Figure 2.2 illustrates the interrelated aspects of the T&E process. This cube relates the test processes, T&E resources, and system maturity and illustrates the cyclical and concurrent nature of T&E. The steps in the test process are repeated as the system matures. The objective is to identify deficiencies as early as possible to avoid the cost and delays inherent in redesign and retest of a more complete,

Figure 2.2
System Maturity, Test Processes, and Resources

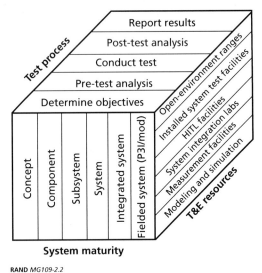

RAND *MG109-2.2*

and therefore more complex, article. Each step of this generic T&E process is described below.[4]

Determine Test Objectives

Specific test objectives are normally derived from the system-level requirements stated in the MNS, the ORD, and the TEMP. These flow down to lower levels through subsystem and component requirements that the contractor derives to support design and testing at each level. Given these performance requirements, test planners specify the article to be tested and the test conditions, methodology, and data required by developing a list of specific test events. An example of a test event would be measurement of key parameters at a specified altitude, speed, attitude, and configuration. The progression of these events is called the *test matrix*. The objective of the events in the test matrix is to gain sufficient knowledge about the performance

---

[4] Air Force Manual 99-110 (U.S. Air Force, 1995) discusses the generic process in more detail.

of the test article under a range of conditions to reduce the risk of proceeding to the next step in development to an acceptable level.

Conduct Pretest Analysis

Before testing begins, the test matrix is evaluated to ensure that it will produce the required data while making efficient use of test resources. Current knowledge, simulations, and other analyses are used to help predict the results to be observed. The responsible test organization develops a detailed test plan, making allowances for unplanned delays.[5] Outside technical experts then review the detailed test plan.

Conduct Test

Prior to conducting the test, the team must select specific test points from the test matrix. In some cases, this may not be a simple progression from basic to more challenging tests because of such external factors as concurrent testing, availability of a properly configured test article, suitable environmental conditions, and required supporting assets. In addition, opportunities for achieving multiple test points during a test event require careful planning and coordination to maximize the productivity of test resources. Efficient and timely execution of a test program becomes a complex balancing act requiring coordination (and often flexibility) among all participants.

Perform Post-Test Analysis

Modern test programs produce large volumes of data, which must be reduced and converted to a useful form for evaluation. Most test activities can produce rapid, "quick-look" reports to give some indication of test outcomes before conducting more-detailed analysis of the test data. Quick-look results can be used during the test itself to verify that suitable data were collected and that the next test can proceed. After the test, the data are analyzed extensively and compared to predicted values. If the predicted values do not match those observed, further analysis is required to determine whether the predictions, test

---

[5] Many test organizations use a test efficiency planning factor of 80 percent to allow for weather, equipment problems, additional testing, or nonavailability of the test article.

conduct, or system under test is responsible. In many cases, a retest must be conducted to validate corrective action.

Report Results

Test results are reported in a variety of ways, depending on the significance of the test and the intended recipients. At the end of DT and OT, formal reports are submitted and retained to provide a record of test program execution and results.

If a valid trial reveals a discrepancy between predicted results and actual system performance, the prediction algorithms are reanalyzed and adjusted as necessary. If a test article deficiency is found, its cause(s) must be isolated, which may require further testing. Once the cause(s) is (are) identified, a design change may be made.

As the type of article being tested progresses from component to subsystem to system, design changes can become increasingly complex and costly because they can affect the operation of other components or performance in other operating regimes. As a result, extensive retests (regression testing) may be required to ensure performance results from previous tests have not been compromised. This situation leads the program manager to a dilemma: Is it better to reduce risk through extensive component and subsystem testing, with known increases in cost, or to save costs through reduced testing, with the potential for much greater disruption and cost if performance deficiencies are discovered later in development. Obviously, the optimal balance between testing and risk depends on the maturity and complexity of the system and on the criticality of potential performance shortfalls.

Selected Test-Related Documentation

This section briefly describes typical documentation that may be useful to an analyst attempting to assess the cost and scope of T&E for a major acquisition program. A number of other reports may provide specific data needed for a thorough analysis, but the reports described below generally provide most information required.

Analysis of Alternatives

The analysis of alternatives (AoA) is a comparison of alternative approaches for meeting the mission needs described in the MNS (or in some cases, the capstone requirements document). AoAs are performed by representatives of the user community, although representatives of the program manager may participate. An AoA is required before Milestone B (or Milestone C, if there is no Milestone B). For the test community, the AoA provides COIs and MOEs to be used in OT.

Operational Requirements Document

The ORD translates the general military requirements contained in the MNS (or capstone requirements document) into specific user and affordability requirements. It also describes how the system will be employed and supported.

To support evolutionary acquisition, ORDs should contain time-phased capabilities. To support the CAIV process, the document should provide threshold and objective values for key performance parameters, to enable trade-offs during system development.

In addition, the ORD requirements should be prioritized, to guide system developers as they make affordability, schedule, and risk-reduction trade-offs. Early versions of ORDs do not necessarily fully define capabilities of subsequent spirals or blocks. Thus, actual operating experience can help refine requirements for subsequent iterations.

The initial ORD is prepared for Milestone B or program initiation. Periodic updates add new details as the development program matures. Because the ORD represents the approved objectives of the users of the system, it is particularly useful for developing OT plans. Test representatives from the services should participate in ORD development to ensure that the requirements are testable.

Test and Evaluation Master Plan

The TEMP provides the overall structure for all parts of the test program and provides the framework for developing detailed test plans. The TEMP describes the development, operational, and live-fire testing; provides a test schedule; and outlines the required resources, including the funding, test articles, test sites and equipment, threat representation, targets, manpower, training, and operational force support. The plan also presents COIs, MOEs, critical technical parameters, and suitability thresholds and objectives derived from the ORD. If there are service-unique test requirements for a joint program, they appear in an appendix.

The TEMP is preceded by an evaluation strategy submitted within six months after Milestone A or program initiation. The TEMP itself is prepared by the program manager and the T&E WIPT, with input from the OT agency on operational testing. The TEMP is first submitted for Milestone B, then is updated before each OT period and for Milestone C, the full-rate-production decision review, and any significant program changes.

Test Resources

The T&E process requires a variety of resources, including contractor and government test facilities, test ranges, manpower, training, flying time, support equipment, threat systems, M&S, instrumentation, communications, range equipment and facilities, data protection and security systems, munitions, and targets. While many of these needs are relatively straightforward to estimate for a given test, the timely availability of appropriate infrastructure to support testing is critical to the success of any test program. Test facilities may be required from any of the following general categories:

- M&S
- measurement facilities
- system integration laboratories
- hardware-in-the-loop facilities
- installed system test facilities (ISTFs)
- open-air ranges.

Figure 2.3
Number of Trials by Various Test Methodologies

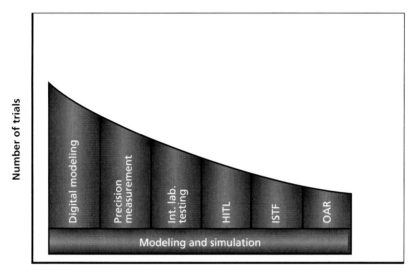

RAND *MG109-2.3*

   These categories can be thought of as representing a spectrum based on level of effort per trial as shown in Figure 2.3. In the figure, the level of effort per trial increases from left to right, as the number of trials decreases. Each is described below.

Modeling and Simulation

Digital computer models are used throughout all phases of system development to simulate and evaluate components, subsystems, and systems. The use of M&S for testing parallels its use in the development process, progressing from relatively generic high-level modeling for evaluating system concepts and alternative architectures to increasingly detailed models as the system design matures and as key system parameters are determined. Typically, as the models mature, they increasingly rely on physical or engineering relationships that can be validated empirically through observation or testing. These component model results can often be aggregated, or the models can be integrated, to produce a system-level model that will predict, with considerable accuracy, the events and conditions that the model architecture and data fully describe. Once validated, these models can be used to predict system behavior under a wide range of conditions at considerable savings in test time and cost.

Digital models consist of algorithms that can be run by providing fixed input parameters, by having the model develop and modify scenarios dynamically within the simulation, or by using an operator in the loop to capture the interaction of human operators with the system in a virtual environment. However, despite experience with the use of validated models, it is unlikely that they will ever fully supplant live testing. Even in areas that are well understood, it is not unusual for live testing to uncover problems that were not apparent in the simulations. Examples include wing-drop effects in the F-18E/F, buffeting of stores in certain flight conditions on a number of programs, and target penetration prediction errors in the Joint Air-to-Surface Standoff Missile (JASSM). These problems seem to occur most frequently when complex interactions are either poorly understood or inadequately modeled.

Models must be verified and validated before use. Verification involves analysis and testing to ensure that the model functions as designed. Validation establishes that the model, when used over some range of conditions, acceptably represents reality. Accreditation is a further step, certifying that the model is appropriate for specific purposes. The development, verification, validation, and accreditation of models is a significant part of early system development and must be planned and budgeted so that it is in place to support T&E.

The benefits of M&S are that they can

- simulate a large number of trials that differ in predictable ways over a short period
- substitute for live testing when safety, availability of appropriate assets or test conditions, environmental factors, or expense make live testing impractical
- support trials in which selected parameters are varied in controlled ways, while others are held constant
- identify system shortcomings before live testing

- identify or refine issues and conditions for live tests to maximize the value of data collected
- increase confidence that the selected live-test data can be interpolated or extrapolated.

Measurement Facilities

These facilities accommodate accurate measurement of parameters of interest under controlled conditions. Such facilities are normally general purpose, although some modification or special equipment may be required for a specific test program. The following are examples of measurement facilities for aircraft and missiles:

- wind tunnels
- propulsion test facilities
- signature-measurement facilities
- environmental measurement facilities
- warhead test facilities.

System Integration Laboratories

These laboratories support integration of hardware and software components in a controlled environment. Components under test can be bench-tested with other simulated components and software. These are generally contractor facilities, although the government may also have such laboratories for testing purposes.

Hardware-in-the-Loop Facilities

Hardware-in-the-loop testing involves exercising actual system or subsystem hardware in a controlled environment using simulated inputs. Hardware-in-the-loop facilities can simulate the systems and threats with which the test article must interact. They are most often part of contractor facilities, although the government maintains them as well.

Installed System Test Facilities

ISTFs allow ground testing of complete systems while they are installed physically or virtually on the host platform. Test facilities may be electronically linked to test data interchange. For instance, an aircraft in a hangar may use its weapon control system to provide commands to an instrumented missile located in another facility. In this way, hardware-in-the-loop facilities may become virtual ISTFs. Examples include anechoic chambers and structural load test facilities.

Open-Air Ranges

Open-air ranges are used for flight testing aircraft and for a variety of guided-weapon tests, from captive carry through live fire. These are the most resource-intensive test facilities, and nearly all are government owned. Table 2.1 lists the primary Air Force and Navy aircraft and missile open air ranges.

Table 2.1
Primary Air Force and Navy Open Air Ranges for Aircraft and Missile Testing

Range	Location
Air Force Flight Test Center	Edwards AFB, California
Air Force 46th Test Wing	Eglin AFB, Florida
	Holloman AFB, New Mexico
Air Force Air Warfare Center	Nellis AFB, Nevada
Air Force Utah Test and Training Range	Hill AFB, Utah
Naval Air Warfare Center—Aircraft Division	Patuxent River, Maryland
Naval Air Warfare Center—Weapons Division	China Lake, California
	Pt. Mugu, California

CHAPTER THREE

Trends in Test and Evaluation

A key objective of this study was to identify changes in the practice of T&E and, to the extent possible, identify their likely effects on the cost of T&E for future aircraft, missiles, and guided munitions. An initial step was to compile a list of those trends we judged most likely to affect T&E cost now and in the immediate future. Offering the results of this process as topics for discussion in advance of our interviews with program and test facility personnel provided some useful structure. We compiled this list from various references, with feedback from Air Force, Navy, and OSD personnel. The rest of this chapter addresses each of the trends we examined.

The trends examined were

- acquisition reform
- M&S
- testing of software-intensive systems
- combined and multiservice testing
- contractor versus government test facilities
- live-fire testing
- warranties.

Acquisition Reform

In defense procurement, *acquisition reform* generally refers to a broad emphasis on eliminating activities that do not add value and ensuring that the remaining activities are as cost effective as possible. A previous RAND report (Lorell, 2001) addresses the general subject of the effects of acquisition reform on cost estimating. In that report, the authors provide a taxonomy of acquisition reform measures:

- reducing regulatory and oversight compliance costs
- adopting commercial-like program structures and management
- using multiyear procurement.

In T&E, all acquisition reform initiatives fall into the first two categories, reducing compliance costs and commercial-like practices. Using the taxonomy described in Lorell (2001), the acquisition reform principles that specifically apply to T&E are

- requirements reform (ORD flexibility)
- contractor design flexibility and configuration control (TSPR)
- commercial insertion (COTS/NDI).

In addition, the ACTD approach, which uses mature technologies to demonstrate new operational capabilities in abbreviated demonstrations, often manifests all these principles.[1] ACTDs are not considered acquisition programs.

The following subsections discuss each of these approaches to acquisition reform in T&E.

ORD Flexibility

The ORD represents the set of mission requirements a new weapon system must meet. It describes the performance that the new system is expected to provide to the operating forces and, throughout the development process, is the authoritative guide for the functional capability the system must attain. The ORD describes these capabilities in terms of key performance parameters, which both provide guidance for the design team and are the basis for the test program.

The ORD development process has, in the past, been criticized as being a "wish list," derived primarily from user desires with rela-

---

[1] OT personnel often conduct or observe ACTD and attempt to provide an assessment of potential military utility.

tively little consideration of the cost of achieving specified levels of performance. One of the fundamental tenets of CAIV is that user "requirements" should allow maximum flexibility, so that degrees of performance can be traded against cost to arrive at a "best value" system, one that recognizes that key performance parameters are not all equally important and that incremental levels of performance were not equally valuable to users.[2]

The CAIV approach involves setting true minimum performance requirements as *thresholds* that must be met, and providing incentives for reaching or exceeding the higher, *objective* or *target* requirements within the overall cost objective. In this way, the contractors, working with the government IPTs, can attempt to maximize value for the government acquisition dollar. This also provides the test community more definitive guidance for assessing operational utility. Most ORDs have supportability and maintainability targets to reduce the tendency to save acquisition cost at the expense of long-term operations and maintenance.

In addition to flexibility in program requirements, acquisition reform also encourages simplification of the ORD to a minimum number of top-level performance measures in an effort to reduce or eliminate the perceived tendency to overspecify. This is intended to promote program flexibility and avoid well-intentioned but inefficient constraints on candidate approaches and designs.

The ORD, in addition to providing guidance for the development of derived specifications for the developers, is also the source of requirements for the test program. Key performance parameters are broken down into *critical issues*, which are the questions T&E must answer. Critical issues must be translated into testable characteristics, often called MOEs, against which specific measures of performance may be observed during testing.

Since the ORD provides the performance requirements for subsequent T&E, it is important to get effective input from the T&E community while the ORD is being drafted. This is to ensure that

---

[2] In contrast, in the earlier design-to-cost approach, the developer's flexibility was limited by a set of requirements that was, for all practical purposes, fixed.

the performance thresholds and objectives are clear and can be verified by testing.

A good example of the importance of the ORD in T&E was provided when DOT&E judged that the Predator unmanned aerial vehicle was not operationally effective or suitable, although it was deployed in Afghanistan and operated with notable success there. The DOT&E evaluation was based on the fact that the system had to be operated well below the requirements set forth in its ORD, the standard against which the tests had to be conducted. If the ORD lacks "crawl, walk, run" performance requirements, system developers and testers must use ultimate performance thresholds as their guidance, even if lower levels of performance would provide useful capability. These phased requirements are expected to become the norm with the DoD emphasis on evolutionary acquisition.

Total System Performance Responsibility

The TSPR concept is an attempt to emulate the commercial research and development environment, in which the company has nearly complete control over (as well as complete responsibility for) the design, development, and testing of a new product. While user requirements, preferences, and budget constraints are significant inputs, the company itself decides on development approaches, specific designs, budget allocations, risk mitigation, and any necessary infrastructure investments. Balancing this autonomy is direct feedback from the market. In the defense environment, the discipline of the market must be created by careful specification and communication of requirements and well-crafted contractual incentives that reward success and penalize poor performance. The presumption is that the possibility of increased profitability will provide incentives for the contractor to be innovative in design and vigilant in reducing unnecessary expense.

TSPR has significant implications for the conduct of T&E. First, it means that the contractor will have primary responsibility for designing and conducting the DT program. While still being required to demonstrate compliance with contract specifications, the contractor has, in many cases, primary responsibility for determining the timing and nature of testing. This may include determining the test facilities to be used, the number of test articles, and the amount of testing. Of course, the test IPT normally contributes to these decisions, since the government must ultimately evaluate the results to determine specification compliance. However, if government representatives request additional tests for risk mitigation or evaluation in a particular facility and if the contractor disagrees, the government may have to provide additional funding through a contract modification.

Second, unless the contract specifies otherwise, TSPR may mean that certain contractor-developed data will not be available to the government test community. For example, the contractor may model airflow as part of the design process. The resulting data might be useful to the government for simulating other events, such as weapon separation. But if certification of the weapons in question are not a part of the original development program, the government may have to recreate the data through its own testing or may have to purchase the data from the contractor separately.

Third, representatives from several programs noted that contractors having configuration control occasionally made modifications that affected interfaces with other systems. Thus, government personnel needed to track contractor-initiated changes closely for effects on related systems.

Finally, TSPR highlights the importance of close cooperation and frequent interaction with government test experts, especially the OT community, so that contractor personnel understand user requirements and operating procedures. Both contractors and government program personnel noted the importance of early identification of potential problems. This is particularly important when documented requirements or specifications have been reduced to foster contractor innovation. Authoritative government input early in the process can save costly redesign and prevent potential contract disputes. Another benefit of frequent interaction is that it improves the familiarity of government personnel with the system, preparing them for evaluating test data.

Commercial Insertions (COTS/NDI)

Although full COTS systems are rare in DoD aircraft and guided-weapon programs, the use of COTS components, when feasible, is becoming increasingly common, both because such components are less expensive than custom military items would be and to take advantage of advances in commercial technologies. A related trend is the use of an existing military system, or NDI, that the U.S. DoD or another country has developed for other applications. Both COTS and NDI have perceived advantages:

- existing production base
- reduced or nonexistent development costs
- low technological risk
- shorter lead times than for new development.

From a testing perspective, we found that using COTS and NDI components does not, in general, significantly reduce system-level testing because the military environment is often quite different from their typical commercial employment. In many cases, normal system-level tests were completed without failures, and COTS components were incorporated into the final design at significant cost savings. In other cases, particularly when entire systems were COTS or NDI, what were assumed to be minor modifications to accommodate DoD requirements became major redesigns, in some cases growing into significant development and test programs.

As part of this study, we interviewed personnel from three aircraft programs that could be considered predominantly NDI: T-45, Joint Primary Aircraft Training System (JPATS), and C-130J. In the T-45 and JPATS cases, the acquisition strategy involved buying U.S.-built versions of existing foreign aircraft with modifications to adapt them to U.S. requirements. In the case of the T-45, the existing British Aerospace (BAe) Hawk was chosen to replace the current Navy intermediate and advanced strike trainers. A firm-fixed-price contract was awarded to McDonnell Douglas to modify the Hawk for aircraft carrier operations. Because it was an existing airframe, little M&S was planned. Initial OT identified major problems in aircraft handling, both in flight and on the ground. Correcting these deficiencies required much more redesign and iterative testing than anticipated. Because neither the contractor nor the government had planned for extensive rework and testing, the program's schedule slipped several times. The aircraft finally completed OT in 1994, ten years after the program entered FSD.

JPATS was the first aircraft to be designated as an acquisition streamlining pilot program. Streamlining acquisition procedures and limiting competition to variants of existing aircraft were expected to reduce the time needed to acquire and field a replacement for Air Force and Navy primary flight trainers. The complete JPATS includes aircraft simulators, training devices, courseware, a training management system, and contractor logistics support, but our discussions with the program office were limited to the aircraft portion of the system.

Source selection included flight evaluation of seven competing aircraft. The Beech (now Raytheon) candidate was selected, and the aircraft reached Milestone II in August 1995. The first two phases of qualification T&E (QT&E) were performed on a prototype aircraft because a production-representative aircraft was not available. One of the acquisition-reform initiatives for JPATS involved relying on Federal Aviation Administration (FAA) certification for portions of the flight-test program. It became apparent, however, that FAA certification is a cooperative effort between the FAA and the requesting contractor to promote basic airworthiness, not to demonstrate compliance with stringent performance specifications. As a result, the planned DoD portion of the flight-test program grew from approximately 50 to 90 percent.

Radio testing is one example of the problems encountered. The FAA commercial standard for radio performance is a ground check, which missed significant dead zones in the antenna pattern. Military testing was required to identify and resolve the problem. Similarly, spin and recovery testing had to be expanded when FAA standards were judged insufficient for training military pilots. The contractor, whose experience with military aircraft testing was limited, did not foresee the additional testing required and the shortfalls it uncovered. Most testing was performed at the contractor's site. As in most TSPR programs, additional government-directed testing required a contract modification.

The C-130J is a contractor-initiated update to the C-130 series of medium-range transport aircraft. Although it retains the exterior and interior dimensions of its predecessors, more than 70 percent of this model is unique, including integrated digital avionics, a redesigned flight station, a new propulsion system, and cargo compartment enhancements (DOT&E, 2000). Allied countries have purchased C-130J variants, and the U.S. Air Force is now procuring it under a commercial acquisition strategy.

The initial T&E approach was to supplement FAA certification with government QT&E in specific military areas of interest. After some delay, a commercial variant that was a conceptual combination of two production aircraft achieved FAA certification. The contractor subsequently decided not to maintain FAA certification. The first seven aircraft had to be modified to bring them into compliance with their model specification, and software problems have required more intensive testing than planned. The contractor planned and executed the DT program, with government DT and OT personnel observing. The program has instituted a CTF to improve coordination of QT&E and OT&E data gathering. The test program is phased to coincide with software revisions. The Air Force conducts its own limited functional software testing, generally on the aircraft.

The primary issues for the C-130J have been numerous software problems and the vulnerabilities demonstrated in live-fire testing. As with T-45 and JPATS, what was intended as a low-risk modification to an existing system under a commercial-type procurement has required far more development time and effort than originally anticipated.

According to the program office, contractor T&E cost data were not available to them because of the nature of the contractor-initiated development effort. As a result, our cost analysis could not include the C-130J.

In summary, these commercial-type procurements yielded the following lessons:

- Both the government and the contractor should clearly understand the differences between the performance capabilities of the system as it exists and those expected in a U.S. military environment. The system requirements review is an important opportunity to define an acceptable system specification clearly.
- FAA certification alone is not sufficient to demonstrate achievement of most military aircraft performance specifications.
- "Best commercial practices" are not an effectively codified set of principles like those of common law or accounting. Because such practices tend to depend on the situation and to be inconsistent from contractor to contractor, they may be inadequate for responsible acceptance of military systems. To be successful, this approach requires extensive communication and, eventually, a mutual understanding between the contractor and customer of specifically what constitutes an acceptable demonstration of system performance.
- Even well-proven commercial products should be tested in a representative military environment. In many cases, they will be found to be as suitable as they are in their civilian applications. In some cases, however, the unique demands of military operations will cause the products to fail, often because of conditions not typically encountered in civilian operation.
- Because of the lack of leverage with the contractor when the military market is a small portion of the contractor's business base, it is critically important to specify performance and test requirements carefully for prospective contractors, to put contractual incentives in place to encourage success, and to ensure enough government insight to recognize impending failure.

Modeling and Simulation

M&S generally refers to the use of computer models to emulate a system to provide insight into its operation without actually operating it. This approach is now used extensively throughout the acquisition process, from evaluating system concepts through operational training.

In T&E, M&S can be used to identify and quantify key system parameters that will become key performance parameters, MOEs, and measures of performance; to model system and component operation; and to evaluate system and component performance. M&S is particularly useful when live testing is not practical because of range, safety, or threat limitations.

M&S is often cited as a transformational advance in T&E technology. The claims that M&S reduces the requirement for live testing, therefore also reducing the expense of the test program, is particularly relevant for our study. Because of the importance of this issue, we specifically highlighted M&S and its potential for cost savings in interviews with test program managers, test activity personnel, and cost analysts.

A wide variety of models and simulations are used in various phases of T&E. Top-down models may use aggregate characteristic data to represent operation of a system in a wide variety of scenarios, including force-level engagements. Engineering or physics-based models are typically based on detailed data and can be used to model components and subsystems or can be integrated to model an entire system. Both types can include interfaces with actual components (hardware-in-the-loop) or human operators (man-in-the-loop). Engineering (or physics-based) models are common in T&E because they often arise from the engineering design activities of the program, or the test activities specially develop them. Their data and relationships are carefully developed and validated, normally starting with a basic capability and expanding over time. Although these models are based on "hard" physical or engineering relationships, they may fall short if the model does not include all relevant parameters, effects, or interactions. In some cases, such elements may be overlooked because they occur infrequently, because their effects are subtle, or because the phenomena to be modeled are not completely understood.

Top-level models, on the other hand, are often used to compare alternatives and set performance thresholds. They are used less often in T&E applications because of the difficulty of objectively verifying their fidelity across the full range of interest.

M&S can take a variety of forms in T&E applications, including

- detailed component testing
- subsystem integration
- system simulation and stimulation
- campaign or force-level (system of systems).

The following are some common advantages of applying M&S in T&E:

- It can overcome or reduce test limitations that are due to range restrictions, safety, or threat interactions.
- It makes it possible to conduct many trials efficiently that involve varying known parameters in predictable ways.
- It provides a means of adjusting or testing parameters for insight, sensitivity analysis, or optimization.
- Once developed and validated, models are extremely valuable for system upgrades or follow-on development efforts.

The benefits can also carry over to other related test programs (e.g., threat simulators) if the models are serviceable and well documented.

Using M&S for testing also has limitations that should be recognized:

- **Level of Fidelity:** Although it may sound obvious, the available models are not always able to reflect real-world outcomes consistently. This generally happens in areas for which the physical phenomena have not been characterized well or in which there are complex interactions that are difficult to predict accurately. Another, more common, problem is limited availability of actual detailed data for model development and calibration. This type of data typically requires specialized instrumentation and repeated tests to collect. Because of the time and expense involved, the collection of adequate data for non–program-specific, high-fidelity modeling seems to be fairly uncommon. A third component of high-fidelity modeling is validating model operation with observations of actual performance. As with data collection, this requires conducting physical tests with sufficient instrumentation to validate model predictions under similar conditions.

- **Range of Applicability:** Models are typically developed and tested over an expected range of input and output values. When models are used outside these ranges, their outputs may become unreliable. The current environment, in which program managers fund M&S development, does not encourage the design or validation of models for conditions outside the program's immediate requirements.

- **Cost of Model Development:** When "affordability" is a major criterion, every significant item of cost in a program budget is scrutinized. In addition to putting the obvious constraints on the total program, program managers must also carefully phase activities to match funding available for a given period. Budget cuts or unexpected problems may require deferring planned activities to later fiscal years. Because of the time it takes to develop and test complex software, effective M&S investments must be made early in the program, well before their intended use. Unfortunately, this is also a time when most programs are not well staffed or funded. Thus, without a strong advocacy, M&S tends to be developed on a limited budget and a constrained schedule. These early limitations may manifest themselves in ways both obvious (schedule slips) and subtle (limits on design flexibility and future expandability).

- **Integration with Other Models and Hardware:** Another often-underestimated aspect of model development is the integration of component- and subsystem-level models into higher-level models. As with individual model development, early funding or schedule constraints may hinder planning for integration of various models. Although the benefits of coordinating the development of lower-level models so that they can easily become part of an effective and flexible system-level model are well recognized, budget, schedule, and occasionally organizational constraints again make this challenging. When a higher-level model is needed and detailed models are not available or sufficiently mature to use as components, a top-down model may be substituted. These models have often been criticized as being less robust because many of the inputs, and sometimes the algorithms themselves, tend to be more conceptual and are rarely subjected to rigorous validation through actual testing.

In summary, M&S has become integral to most development programs. It supports testing of components and systems when live testing is not economically or practically feasible. It often allows alternative designs to be tested and iterated before commitment makes changes prohibitively expensive. Comprehensive M&S programs may also contribute to other areas, such as development of tactics, training, and future system enhancements. There is general agreement that M&S has reduced the amount of live testing that otherwise would have been required, although other test activities tend to mask these savings. It also had the affect of making most live testing more focused and robust than in the past. The Defense Science Board, in its 1999 review of T&E, found that

> Claims of substantial program cost savings attributable to the increased use of M&S, with a concomitant reduction in testing, cannot be verified. ... a White Paper prepared by the AIAA [American Institute of Aeronautics and Astronautics] Flight Test Technical Committee (FTTC) in late 1988 entitled, "Seeking the Proper Balance between Simulation and Flight Test," states "the members of the FFTC are unaware of any study that has supported the claim of substantial program cost savings realized by a significant expansion of the use of M&S with a concomitant reduction in testing."

Similarly, the Air Force Scientific Advisory Board (1998) concluded that

> Evolutionary improvements in the use of M&S enhance T&E and may eventually reduce the costs; however, they cannot be expected to become a cost-effective or technically sufficient

replacement for most physical testing of air vehicles in the fore-seeable future.

As in system design, M&S has increased the efficiency of system testing. In such areas as aerodynamic modeling, it has reduced the wind-tunnel hours required. The Air Force SEEK EAGLE program can now do computerized fit checks for stores compatibility. In fact, a number of test personnel admitted to us that the M&S capabilities for typical programs are being used to such an extent that they are hard pressed to keep up with the demands for capability and fidelity.

The goal of many in the test community is for M&S to be the primary iterative test method, with live tests used for validation. Today, M&S is routinely run in predictive mode before live tests and is modified as required to reflect observed outcomes.

Testing of Software-Intensive Systems

Most current aircraft and guided-weapon programs can be classified as "software intensive." For a variety of reasons, much of the func-tionality of modern weapon systems is actually implemented in soft-ware. This has advantages for upgradeability and evolutionary devel-opment, as well as for improved availability and supportability of common hardware components. However, developing and testing complex software presents some of the most difficult challenges in system development, particularly when a large number of functions and a high level of integration with other hardware and software strain the ability of current development processes and tools.

Methods of improving software testing are being investigated and implemented across a wide range of government and commercial programs. Although it is beyond the scope of this report to address this issue in detail, we can summarize some of the approaches relevant to aircraft and guided-weapon software testing.

A common predictor of future software development problems is incomplete, unclear, or changing requirements. Although this problem is well recognized, many programs continue to suffer from poor definition of software requirements. In addition to leading to faulty software, poor requirement definition compromises test activities by making it more difficult to design and execute effective software testing; by leading to late discovery of problems and delays in testing while code is fixed; and by requiring additional time and effort for regression testing.

Virtually all modern software development involves some form of structured development approach. For large or complex systems, this generally involves some form of evolutionary or "spiral" development, which introduces additional software functionality in sequential releases or "blocks." These control the amount of new code and functionality that must be tested in each release, in theory building each subsequent increment on relatively stable, well-tested modules. While this approach may produce higher-quality software, several programs noted that it complicates system-level test planning by restricting testing to the capabilities implemented in the current release and by possibly increasing the amount of regression testing required.

In missile programs, the use of "captive carry" flights to test missile components on an aircraft simulating the missile's flight is well established. Similarly, on several larger aircraft programs, flying test beds were used for testing hardware, software, and the integration of the two. In general, the programs that used flying test beds judged them to be very effective. They allowed early DT in conditions closely simulating the operational environment, without the problems and expense of testing immature systems and software on the actual developmental aircraft. For programs with significant software but without a flying test bed, one or several robust ground-based integration laboratories were a virtual necessity.

Since testing accounts for a significant portion of the software development effort, software can be designed for improved testability by considering and accommodating test requirements throughout the development process. Since manual test design is labor intensive and error prone, automated generation of test cases and automated, adaptive test algorithms can both improve the quality of testing and reduce the time required.

In summary, the challenges software-intensive systems present will continue to grow with the introduction of more complex systems and with the more-stringent interoperability requirements of the "system of systems" approach to defense acquisition. With this increase in complexity, costs of testing software-intensive systems can be expected to grow unless testing efficiency can somehow be improved to offset the complexity increases. The common goal of all the testing approaches described above is to identify and correct problems as early as possible, minimizing those discovered in flight testing.

Combined and Multiservice Testing

Combined testing generally refers to the integration of contractor and government DT and government OT personnel on a single test team (often referred to as a CTF). Previously, the contractor would generally perform much of the DT, with dedicated government test periods at the end of each major phase. Similarly, government operational testers had a largely independent test plan for OT. Now, with integrated test teams or a CTF, government DT and OT personnel are involved from the early planning stages through the completion of all major test activities. Our contacts felt that the early involvement of OT personnel in DT saved both costs and schedule for the program.

The statutory requirement for an independent organization to do operational evaluation at the system level is now interpreted as requiring only independent analysis of representative test data. Air Force government-only testing (IOT&E) and Navy OPEVAL are normally conducted in dedicated tests that the OT organization performs, augmented, if necessary, by other service assets. With the exception of any contractor logistics and maintenance support that would be a normal part of the operation of the system, neither the contractor nor the government development organization participates directly.

In general, the advantages of combined test teams are

- elimination of redundant test activities
- early identification of issues and problems

- effective communication on potential efficiencies and work-arounds throughout the test program.

Nearly all interviews indicated that program office, contractor, and OT personnel see early involvement as positive and useful, making such comments as "tremendous benefits in terms of schedule and cost." Ideally, the operational testers are directly integrated into the CTF as full participants throughout the test program. The only negatives mentioned were some potential for an adversarial relationship to develop between the IPT representing the acquisition function (which is therefore trying to field the system in the minimum time and cost) and the operational testers representing end users and maintainers (who are attempting to maximize utility once the system has been fielded). There was also some concern in the OT community that informal opinions rendered early in the test program could be perceived as definitive commitments by the OT activity. A consistent concern on the part of both Air Force and Navy test communities was the perceived shortage of qualified OT personnel. This often limited their ability to support extensive early involvement, particularly for smaller, less-visible programs. In some cases, AFOTEC has hired support contractors to fill specialized slots for which Air Force personnel are not available.

One of the contractor test managers observed that, because of the degree of management integration in modern test programs, cost analysts must be cautious about trying to isolate discrete elements of a test program artificially, since the objective is to conduct many previously separate activities concurrently and thus gain synergy from the integrated activities.

Multiservice testing involves systems that more than one service will use. In these cases, which are becoming the norm for air-launched weapons and, to a lesser extent, for aircraft, a lead service is designated and has primary responsibility for executing the test program. In general, requirements that are unique to one of the participating services are normally tested by that service as part of the overall test program. The cost implications of multiservice testing depend on the degree of commonality of requirements and configurations among the participating services.

Contractor Versus Government Test Facilities

Another question we posed in our interviews was whether there had been any consistent trend toward the use of either government or contractor test facilities. This issue was raised because of economic pressures on both the government and contractors—on the government to close duplicative or low usage facilities and on the contractors to reduce fixed infrastructure costs. The dilemma the government faces is the conflict between maintaining facilities that are perceived as providing critical or unique test capabilities and following a fundamental principle of acquisition reform, that the program manager must have maximum flexibility to optimize his or her test program. The first goal would lead to a policy mandating use of government facilities for testing, while the second goal would leave the program manager free to choose where to test. We were specifically interested in whether the recent trend in some programs of giving the contractor wide latitude in how and where the test program will be executed resulted in significant shifts in test facilities.

Not surprisingly, the consensus from both the government and contractor personnel was that the contractors would generally prefer to do as much testing in their own facilities as possible. They feel that this gives them greater control over cost and schedule by minimizing travel costs, facilitating communication, and reducing charges for infrastructure that does not add value to the product. Of course, few contractors have open-air test ranges or certain specialized facilities that they do not commonly use as part of their normal operations. As a result, nearly all open-air testing for aircraft and guided weapons, and for such specialized operations as climatic and electromagnetic effects, electronic warfare (EW), carrier suitability, etc., is done at government ranges and facilities. Contractor facilities are generally used for M&S and system integration laboratories, since these are also used for development activities. In general, contractors have the capability to do most component and subsystem testing in their own facilities. Hardware-in-the-loop testing can be done in either government or contractor facilities, depending on circumstances and availability.

While there was general agreement that the major government test facilities are essential for executing the required test programs, some contractor personnel expressed varying levels of frustration in their dealings with the government test organizations. In programs with aggressive affordability goals, there was a concern that some government test organizations were not as focused on controlling the costs and schedule of the test program as other members of the test team were. The respondents felt some practices at the test ranges were overly conservative and caused unnecessary costs and delays. In some cases, delays resulted from chronic understaffing or rigid procedures with little provision for flexibility (particularly in areas perceived as involving safety). A representative of one contractor noted that its criteria for selecting among the available test facilities on a recent program were, in order,

- cost
- responsiveness
- past performance
- capability.

Another contractor representative noted that the government ranges tended to be "overfacilitized but undermodernized," with extensive (and often unique) infrastructures to support and limited funds for investment in modernizing test equipment and facilities. These issues are of increasing importance with TSPR contracts, in which contractors are attempting to perform within what effectively become fixed test budgets and schedules.

One of the challenges government test facilities face is that they must recoup a high percentage of their operating costs through user fees. All the ranges we visited spend considerable time and effort to set these fees by projecting costs and usage for the coming fiscal year. Although this is integral to financial management in industry, it is more challenging in a government environment, in which managers have limited control over some parts of their cost structure. In addition, although test ranges are, in theory, involved in test planning for all major programs, there is a significant amount of schedule "churn." For example, at one range, half the actual testing for the year reportedly was not included in the original schedule used for budgeting. Despite these difficulties, both government and contractor program personnel said significant unanticipated changes in range use rates were fairly uncommon. When they did occur, the result was usually reduction or deferral of the planned testing.

Given the occasional schedule turmoil, test article or range equipment problems, weather delays, and limitations on range capacity, surprisingly few programs complained of getting "bumped" because of higher-priority users. This could have been due, in part, to the fact that most of the programs in the study were relatively large or high priority. We did hear of range availability delays of up to six months for one FMS program.

Live-Fire Testing

Full-scale system-level live-fire testing (or a formal waiver certifying that it would be unreasonably expensive or impractical and describing proposed approaches for assessing vulnerability, survivability, and lethality) has been a congressionally mandated requirement since November 1986. Such testing has therefore been a part of all covered programs since then. However, we found that the costs were often combined with other DT activities.

Although the requirements for live-fire testing, especially for modifications of previously fielded systems, are often hotly debated at the service headquarters level, most program personnel seemed to consider the testing to be a "fact of life" requirement and planned accordingly. This could be because live-fire test plans and any waiver requests must be submitted and approved early in the program (before Milestone II or B), and are therefore integral to all subsequent test planning. In the case of modification programs, the cost of live-fire testing varies depending on whether the modifications are likely to effect vulnerability, survivability, or lethality. Current programs commonly make extensive use of M&S for vulnerability, survivability, and lethality analysis, as well as for test design, to maximize the cost-effectiveness of testing.

Warranties

None of the government or industry test personnel interviewed indicated that warranties significantly changed the T&E process or costs. The T&E scope was set independently of warranty considerations.

Cost Trends

In the preceding subsections, we examined trends in various aspects of the T&E process. A key question remains: Have these initiatives, in the aggregate, had the effect of reducing the overall cost of T&E?

Although it is difficult, if not impossible, to quantify the cost effects of each of these in isolation, we did attempt to determine the overall direction of T&E costs since the early 1970s. Although test programs clearly differ in content, both individually and by system type, it is at least apparent that costs are not trending downward. We also examined the possibility that test durations might be shorter. Figure 3.1 shows that, if anything, they are lengthening.

Have all the changes to the T&E process described above really been ineffective? based on the interviews and some circumstantial evidence, it appears that any net savings due to increased efficiencies in testing are being offset by other trends.

Improvements in testing efficiency and effectiveness are difficult to isolate and quantify from the available data. Nearly all the test personnel in government and industry commented on the increased productivity made possible by advances in M&S data collection systems and approaches and by the reduction of redundant testing between contractor and government DT activities and between the DT and OT test communities. While it is difficult to assess productivity

Figure 3.1
Flight-Test Duration of Aircraft Development Programs Over Time

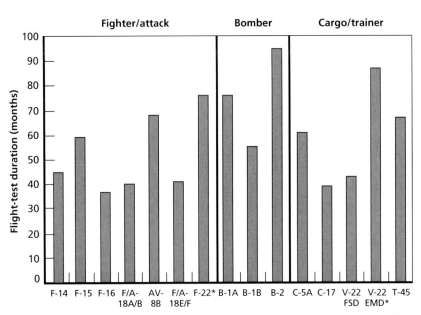

NOTES: Measured from first flight to end of DT. By comparison, the Boeing 777 flight-test program lasted only 11 months, during which nine test aircraft provided 69 aircraft months of testing and logged approximately 7,000 flight hours. Of course, commercial aircraft do not require testing in the range of mission requirements and flight regimes typical of military aircraft.
* Program in progress.
RAND *MG109-3.1*

improvements objectively, given the lack of consistent test activity data over time, we did find comparisons in some areas. While these should be considered to be examples rather than a definitive sample, they tend to confirm much of the qualitative feedback on testing trends we received from government and industry personnel.

The proliferation of digital systems on modern aircraft has shifted much functionality (and complexity) from hardware to software. In some ways, this has actually improved testability by shifting test activities from mechanical, electrical, and hydraulic systems to software and firmware, which not only can be tested more efficiently in ground facilities but can themselves facilitate testing and data collection. For example, aircraft fatigue testing previously involved designing and building special test equipment to provide required inputs (movements, deflections, and loads) to the test article. With digital flight controls, precise inputs can be commanded through the flight control system itself. Some aspects of digital flight control and avionics performance can be monitored by recording data directly from the digital data bus without the weight, space, and expense of dedicated test sensors.

The amount of test data recorded is one indicator of the increased scope of testing. Test equipment installed on the F/A-18A/B would typically generate approximately 256 kilobytes of data per flight. By comparison, the Joint Strike Fighter (JSF) is projected to provide 3 to 4 gigabytes, much of it through the fighter's own data bus. A related improvement, one Boeing strongly emphasized, is that test engineers participate in system design to address testability and test instrumentation concerns. This allows for efficient installation of provisions for instrumentation and wiring during design and test article manufacturing rather than trying to do so later, on an ad hoc basis.

However, despite these advances in the conducting of tests, other changes appear to have offset any potential net savings. Most of these changes can be assigned to one or more of the following categories:

- system complexity
- higher test standards
- increased test charges to programs.

Continuing advances in technology have translated into system capabilities unimagined a generation ago. The growth in capability translates, at least indirectly, into increased test complexity. Figure 3.2 shows the relative growth in the amount of flight-test data collected for three representative fighter aircraft developed since the early 1970s. Although it is normally simpler to collect digital data than to collect the corresponding analog data, the figure makes it clear that the amount of data to be analyzed has greatly increased. Table 3.1 highlights the effects of the advanced capabilities of the F-22 on testing.

Figure 3.2
Growth of Flight-Test Data Collected as Indicated by Number of Sensors or Measurement Points

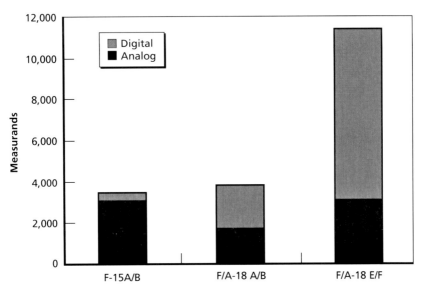

SOURCE: Boeing.
NOTE: *Measurands* are the number of sensors or measurement points.
RAND *MG109-3.2*

Table 3.1
How Advanced Capabilities Affect Testing

Feature	Requires
Reduced-signature design	Maintaining several test aircraft in low observability (LO) configuration
	Obtaining and scheduling unique LO test assets
	Managing security
Internal weapon carriage	Additional wind-tunnel characterization of the flow field with the bay open
	An additional flight-test configuration (doors open) for performance and flying qualities
Sustained supersonic cruise	Reducing test time with chase aircraft
	Adding tanker support
	Increasing the use of supersonic test airspace

Table 3.1—continued

Feature	Requires
Thrust vectoring	Special ground-test fixtures to control vectored exhaust gases
	Multiaxis force and moment instrumentation for measuring thrust
	Ground and in-flight performance testing
	Expanded flying and handling quality testing
	Failure modes and effects testing, particularly with respect to asymmetric actuation
Integrated avionics	Additional EMI and EMC testing
	Comprehensive ground and air testing of integrated system modes
	Collecting data on system timelines and their effects on system performance
Sensor fusion	High-density, multispectral, integrated, enhanced-fidelity target and threat simulation
	Comprehensive integrated ground-test facilities
Highly integrated, wide-field-of-regard sensors	Multiple threat and target simulators with high update rates operating concurrently and having a large field of view
Tailored countermeasures	A target platform with a representative signature
	Air and ground threats that appropriately stimulate the system to determine countermeasure effectiveness
Integrated maintenance information system and technical order data	First-of-kind DT/OT evaluations and assessments (for software-intensive, paperless systems)
AFMSS/mission support element	A higher level of integration in the testing activity, because of the higher level of integration of these systems in the weapon system

SOURCE: F-22 TEMP, 1999.

In addition to the obvious increases in the complexity of the systems under test, there is also some indication that the standards to which current systems are tested are higher than those for legacy aircraft and missiles. Table 3.2 compares the metrics from fatigue testing of the F/A-18 A/B (developed in the late 1970s) with those for the F/A-18 E/F (developed in the mid-1990s). (Despite the nearly twelvefold increase in the data collected, the labor required dropped by approximately half.)

Similarly, the F/A-18 A/B was certified for two weapon configurations during DT; the F/A-18 E/F DT certified 29. Current plans

Table 3.2
Comparison of F/A-18 Fatigue Testing

Metric	F/A-18A/B	F/A-18E/F
Strain gauges at start[a]	615	1,643
Deflections at start[b]	18	89
Data channels[c]	132	1,560

SOURCE: Boeing.
[a]Represents the amount of instrumentation.
[b]Represents the number of cycles.
[c]Represents the data volume.

for the JSF are to certify more than twice as many weapons as the
F/A-18E/F. Current projections show that the JSF will have roughly
three times the 2.1 million lines of code projected for the F-22 air-
vehicle software.

Overlaid on the increasing complexity and scope of test programs is the increase in the fees for using DoD test facilities. As
Figure 3.3 shows, user fees pay a significant proportion of the cost of
operating test facilities. This is consistent with the fee-for-service shift
many DoD activities have made. While this does not change DoD's
overall cost, spreading fixed costs among fewer users increases the
costs for individual programs, unless infrastructure costs can be
reduced proportionately.

Figure 3.3
Comparison of Institutional and User Funding for Major Air Force and Navy
Aircraft and Weapon Test Centers

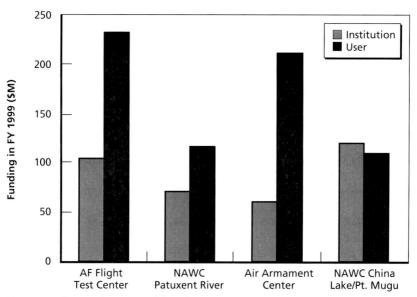

SOURCE: Defense Science Board (1999).

CHAPTER FOUR

Test and Evaluation Cost Data

This chapter discusses T&E cost data, addressing the following in particular:

- T&E cost-element definitions
- how contractors and the government develop and report costs
- the cost data sources for this monograph
- cost data caveats
- how this analysis aggregated the cost data.

Cost Element Definitions

In any study of this type, it is important to define clearly the costs that are being analyzed and those that are being excluded. Military Handbook 881 (MIL-HNBK-881) provides guidance on developing and using work breakdown structures (WBSs) in the development and acquisition phases of DoD programs. Although the definitions from the handbook we present below are largely verbatim, we have condensed them and have omitted many of the examples in the handbook. See Appendix C for the actual Section H.3.3 (ST&E) from MIL-HNBK-881.

System Test and Evaluation

ST&E is the use of prototype, production, or specifically fabricated hardware or software to obtain or validate engineering data on the performance of the system during the development phase of the program. ST&E includes detailed planning, conduct, support, data reduction and preparation of reports from such testing, and all hardware and software items that are consumed or planned to be consumed during such testing. ST&E also includes all effort associated with the design and production of models, specimens, fixtures, and instrumentation in support of the system-level test program.

This category does not include test articles that are complete units; these are funded in the appropriate hardware element. So, for example, the cost of manufacturing flight-test air vehicles should be included in the Air Vehicle WBS element, while static, fatigue, and drop test articles are included in ST&E.

The handbook divides ST&E into five main elements:

- development T&E (DT&E)
- OT&E
- mock-ups
- T&E support
- test facilities.

Development Test and Evaluation

DT&E is planned and conducted and/or monitored by the DoD agency developing the system. The ultimate purpose of DT&E is to demonstrate that the development process is complete and that the system meets specifications. The outputs are used throughout development to support performance evaluation, trade-off studies, risk analysis, and assessments of potential operational utility. DT&E includes wind tunnel, static, drop, and fatigue tests; integration ground tests; test bed aircraft and associated support; qualification T&E, developmental flight tests, test instrumentation, and avionics testing. Table 4.1 shows representative contractor activities normally included in DT&E for aircraft and guided weapons.

Operational Test and Evaluation

OT&E is the T&E that agencies other than the developing command conduct to assess the prospective system's military utility, operational

Table 4.1
Representative Contractor DT&E Activities for Aircraft and Guided Weapons

Aircraft	Guided Weapons
System test requirements and planning	System test requirements and planning
M&S	M&S
Wind-tunnel tests	Wind-tunnel tests
Static article and test	Structural tests
Fatigue article and test	Environmental tests
Drop article and test	Special test articles
Subsystem ground tests	Other ground tests
Avionics integration tests	Flight-test support
rmament and weapon delivery integration tests	Test aircraft preparation
Contractor flight test	Telemetry kits
Special testing	Targets
Other T&E	Other T&E
T&E support	T&E support

effectiveness, operational suitability, and logistics supportability. OT&E includes any contractor support used during this phase of testing.

Mock-Ups

Mock-ups encompass the design engineering and production of system or subsystem mock ups that have special contractual or engineering significance or that are not required solely for conducting either DT&E or OT&E.

The reported costs for mock-ups have been only a small part of total ST&E for recent programs, and the relative costs of mock-ups and the share of total ST&E costs appears to be declining over time. This decline may be due to advances in computer modeling that reduce the need for elaborate physical mock-ups. It may also be that most mock-up costs tend to be reported under DT&E rather than separately.

Test and Evaluation Support

T&E support includes the effort necessary to operate and maintain, during T&E, systems and subsystems that are not consumed during the testing phase and are not allocated to a specific phase of testing. It includes, for example, spares, repair parts, repair of reparables, warehousing and distribution of spares and repair parts, test and support equipment, test bed vehicles, drones, surveillance aircraft, contractor technical support, etc. We found that the content of T&E support varies considerably across programs.

Test Facilities

Test facilities here are the special facilities required to perform the DT necessary to prove the design and reliability of the system or subsystem. These facilities include white rooms and test chambers but exclude brick-and-mortar facilities identified as industrial.

In general, program offices fund only the test facilities that are unique to their program. Once a facility is available, succeeding programs may arrange to use the facility and thus pay some of the variable costs. It can be difficult to predict the program office's share of such costs because the required test facilities may be available as part of the existing infrastructure; may be funded by direct test infrastructure funding; or may require funding from the program office, either alone or in conjunction with other users. It is noteworthy that the B-2 and F-22—programs with unprecedented levels of sophisticated avionics, LO technology, and high levels of security—dedicated considerable portions of their T&E budgets to paying for special facilities that were not available as part of the existing test infrastructure.

Cost Collection and Reporting

The government program manager is responsible for developing the new system and decides how to acquire each of its elements. At the outset of a program, the program manager defines a WBS that represents the system and supporting activities in a product-oriented hierarchy consisting of hardware, software, facilities, data, services, and other work tasks. This hierarchical structure completely defines the system and the work to be done to develop and produce it. MIL-HDBK-881 contains generic three-level WBSs, organized by commodity type (including aircraft and missiles), to provide program managers a starting point for defining the WBS for a specific program. The program manager usually places certain WBS elements on contract and requires the contractor to report costs according to the defined WBS.

Contractor Costs

Contractors report their costs to the government using a variety of standard reports, primarily the cost performance report (CPR) and variations of the contractor cost data report (CCDR). The source of the cost data for these reports is the contractor's accounting system. Since the cost account structure of most contractors' accounting systems will not match the approved program WBS, the costs are allocated or grouped into the government-approved program WBS for reporting purposes (providing a *crosswalk* between the accounting system and the cost report).

The DoD has used this well-defined and accepted WBS structure, and standardized cost reports tied to it, for decades. This has resulted in a large body of historical contractor costs that are intended to be comparable within a given commodity type. The degree of comparability depends on how well the crosswalk has been done, which itself depends on the degree of difference between the contractor's account structure and the program WBS and on the amount of effort that has been devoted to reconciling them. Because of the long-standing requirement for cost reporting and the routine use of the data, for both management of current programs and analysis of future programs, the system generally produces a useful high-level picture of program costs. The fidelity and/or consistency of cost reporting tends to decrease at lower levels of the WBS, for smaller programs, with the extensive use of integrated product (or process) teams, and when commercial practices are used as a justification for reducing or eliminating contractor cost reporting.

Government Costs

In contrast to the long-standing and consistent method for capturing and reporting contractor costs associated with weapon systems, the methods used to capture and report government costs associated with weapon systems vary across organizations. Government financial reporting systems vary in their accuracy, categorization, comprehensiveness, level of detail, and availability. We found no central repository of government test costs at the military service or systems command levels. Rather, costs were captured, reported, and stored at the organizations directly responsible for testing and/or in the individual program offices. The following paragraphs describe the funding sources and rate structures for test activities and how government costs are captured and reported at the organizations responsible for testing, based on our visits to several such organizations.

The rules governing the funding of test organizations, the processes by which they are funded, and their rate structures are complex, and the details are beyond the scope of this monograph. These paragraphs provide a context from the perspective of users of test facilities. Test organizations receive funding from a number of sources. For the purposes of this report, the funding sources fall into two categories: users of the facilities and all other sources. The primary users of the facilities include DoD program managers; contractors working on DoD contracts; foreign military sales (FMS) organizations; and, to a lesser extent, non-DoD contractors, state and local governments, and private parties. In addition to users, the other sources of funding include appropriated funds classed as institutional or "direct budget authority." These funds are intended for maintaining and upgrading the general-purpose test infrastructure.

All the test facilities we visited and from which we collected costs are part of the Major Range and Test Facility Base (MRTFB). The MRTFB comprises major test activities regarded as national assets that receive institutional funding from DoD. The test activities are required to use a uniform reimbursable funding policy, under which users are charged for direct costs, such as labor, materials, equipment, and supplies, and pay a proportional share of equipment maintenance costs related to their use. In practice, the direct costs and proportional share of costs charged to users are affected by the requirement for the test activity to break even. The test activity must balance its costs against institutional funding and customer funding and must adjust the rates it charges customers so that the costs they incur and the funding balance.[1] The test facilities expend considerable effort avoiding unplanned rate increases because of the obvious problems for users' test budgets. Although there are differences in the way Navy and Air Force test facilities are funded, the funding policies and practices at the MRTFB facilities we visited were similar because they are all governed by the same DoD financial regulations. It should be noted that these funding practices apply to DoD facilities and not to the National Aeronautics and Space Administration or private facilities that DoD occasionally uses for testing.

In summary, the user organization pays for the costs of labor and materials related to the tests conducted at the facility and pays a proportional share of other test-related costs. The user's rates are affected by how much other business is at the facility that year.

The funding practices for OT vary more significantly between the Navy and Air Force. For the Air Force, AFOTEC has its own program element and pays for significant items, including range costs, rental of equipment, contractor support, special supplies and equipment for data reduction, civilian pay, per diem and travel, supplies and equipment, and pretest planning. For the Navy, OPTEVFOR pays for its own personnel and travel; the system program manager pays for everything else. After OPTEVFOR develops a program OT budget, the Navy program manager sends funds to OPTEVFOR for its contracting office to arrange for range time, assets, additional contractor support, etc. This difference in funding, although a relatively small part of the total T&E costs for a weapon system, would tend to make OT costs for a Navy program somewhat higher than for a similar Air Force program, all else being equal.

The process of estimating government test costs for a program begins when the program office identifies the need for testing. In the Air Force, the SPO prepares a program introduction document (PID) that identifies the system that needs to be tested, the test services and

---

[1] We attempted to determine whether there had been significant shifts of costs to users to compensate for shortfalls in institutional funding at the test centers but, because of the lack of consistent data over time, could draw no conclusions.

test equipment required, the starting and ending dates, and other related information. The program office sends the PID to the organization that will conduct the testing. The test organization responds to the PID with a statement of capability (SOC). The SOC confirms that the test organization has the resources to perform the testing and contains a summary of proposed test events and resources, a schedule, and costs. The SOC also provides an assessment of technical, schedule, cost, and programmatic risks. The program office consents to the terms in the SOC by providing funding and written confirmation to the test organization.

The test organizations we visited track budgeted funding, obligations, and expenditures, generally by organization or function. However, in contrast with the tracking of contractor costs, there was no consistent product or output or related WBS for government costs across programs, so comparison across programs below the level of total government costs is not currently practical.

The Navy's process is not as formal as that of the Air Force. Financial obligations and expenditures are typically tracked and recorded in the program offices that originate the funds and in the test activities. However, the Naval Air Systems Command (NAVAIR) cost analysis group's attempts to extract T&E cost data for this study from the Navy's Standard Accounting and Reporting System (STARS) were generally unsuccessful.

As in the Air Force, the lack of a standardized WBS for government costs results in a wide disparity in the level of detail of cost data. For example, in some Navy programs, STARS could identify only total funds by appropriation by fiscal year. Below the appropriation level, identifying the amount spent on system T&E was impossible.

A more significant difficulty is that many government organizations dispose of or archive the cost data, so that it is often unavailable a few years after expenditure. During budget execution, most programs can track their expenditures by activity and specific tasks. However, after the funds have been expended, this information does not appear to be systematically retained for future analysis. In some cases it is archived but, it is, for all practical purposes, unavailable. In other cases, it is simply disposed of. The notable exception was the

Earned Value Cost Analysis System in the 46th Test Wing at Eglin Air Force Base, which is used for both management and analysis. Costs for OT were, however, available for both services from their OT agencies.

Cost Data Sources

We collected costs for use in this report from a variety of sources. For contractor costs, we generally collected and used CCDRs. These reports have the advantage of providing all the contractor's costs on a contract in an approved, uniform format. We collected government costs either from the cognizant program office or from the test organizations involved, using whatever records were available to them. Using the program office as the source of cost data has the advantage that these costs should include all expenditures of program funding, regardless of executing activity. The disadvantage is that these costs are not consistently collected or retained. Information from test activities is generally more detailed but may exclude funding and effort on the program at another test activity.

Table 4.2 lists the sources from which we obtained contractor and government costs for the aircraft programs. Similarly, Table 4.3 lists our sources for contractor and government costs for the guided-

Table 4.2
Aircraft Cost Data Sources

	Contractor Costs		Government Costs		
	CCDR/CPR	**Program Office Records**	**Accounting System (STARS)**	**Program Office Records**	**Test Activity Records**
B-1 CMUP	X			X	X
B-2	X	X		X	X
C-17	X			X	X
F/A-18E/F	X			X	X
F-22	X			X	X
T-45	X		X		X
V-22	X			X	X

Table 4.3
Guided-Weapon Cost Data Sources

	Contractor Costs		Government Costs	
	CCDR/CPR	Program Office Records	Program Office Records	Test Activity Records
AMRAAM FSD	X	X	X	X
AMRAAM Ph. 1		X	X	X
AMRAAM Ph. 2		X	X	X
AMRAAM Ph. 3		X	X	X
AIM-9X	X	X	X	X
JASSM	X	X	X	
JDAM	X			X
JSOW	X	X	X	X
SFW	X		X	X
SLAM-ER	X		X	
WCMD		X	X	

weapon programs. In addition to these programs, our CER development data set included data from selected legacy programs.

Cost Data Caveats

We made every effort to use credible sources of cost information and to check the information for accuracy and completeness. However, because of the different sources of costs and their varying quality and reliability, we have less confidence in some data points than in others. In addition, estimators who are using this information need to be aware that several of the programs have peculiarities that affect their costs. We therefore offer the following caveats:

- The B-1A program was cancelled before production, but flight testing continued at a low level for several years before the B-1B program began. Flight-test costs for the B-1A are higher than those of other programs for the number of flights, flight hours, and flight months and may represent additional effort in the interim period between the formal development programs.

- The B-1B airframe was derived from that of the B-1A, so the B-1B test program had less ground testing than would a normal development program. For CER development, we combined the costs for both test programs.
- The F-22 program was in progress as of this writing and is included for information only. It was not used in any quantitative analysis because the content and cost of the remainder of the flight-test program were uncertain. The costs shown were developed from the estimate at completion (EAC) for flight test from the November 2001 CPR that the program office provided, plus the EAC for all other test elements from the March 2001 CCDR.
- The proportion of ST&E costs relative to the total development program is much smaller for the F-22 program than it is for any other aircraft in the data set. There are at least two probable reasons. One is that the program was in progress as of this writing, and the final test cost may be considerably higher than the current estimate at completion. The other possibility is related to the F-22 program's IPT organization and time-charging practices. Contractor personnel who are primarily assigned to a nontest IPT but who work on test tasks charge their time to their original (nontest) IPT. This understates the effort expended on testing relative to other programs.
- As Chapter Three discusses, the T-45 test program was executed primarily by the contractor, with short periods of dedicated government testing. This minimal level of government involvement contrasts with the other programs for which we have government costs, in which the government had a larger, continuous presence and a more-active role. The government test costs for the T-45 program are a much lower proportion of total DT costs than for other programs in the data set. Furthermore, the source of these costs is the Navy's STARS, which may not fully reflect all government test costs.
- The V-22 EMD test program was in progress as of this writing and is included for information only. The content and cost of the remainder of the program are uncertain. We used the EAC

from the December 2001 CCDR. By that date, the originally planned test program was virtually complete. The current test program is event-driven and could continue for an additional 30 months.

- The AIM-9X DT program was largely complete as of this writing. We used an estimate at completion through fiscal year 2002 provided by the program office.

- Advanced Medium-Range Air-to-Air Missile (AMRAAM) Phase 3 is being executed on a TSPR contract, which gives the contractor more latitude and responsibility in executing the contract. This contracting arrangement results in higher contractor test costs relative to the total cost because the contractor arranges and pays the government entities for testing and costs that the government would traditionally pay directly.

- The Joint Standoff Weapon (JSOW) Baseline (AGM-154A) was used for analysis. An additional JSOW variant (AGM-154B) was added to the original JSOW development contract as a concurrent modification; it has not gone into production. Program cost reporting combined both variants, which made segregating Baseline costs problematic. Although NAVAIR and the Air Armament Center attempted to identify Baseline costs for us, the cost data should be used with caution.

Many programs, especially missile programs, have test-related efforts that are not reported as part of system T&E. The reporting of test-related costs can vary from one contract to another, even for modifications to the same basic missile by the same contractor. For example, simulation costs were reported in ST&E in one AMRAAM modification program, but in different elements in the other AMRAAM modification programs. In general, the distinction between component-level testing and system-level testing can be unclear, and contractors use their best judgment in allocating costs. Although we are aware that there are inconsistencies among reported costs, we have not tried to adjust the costs as reported because we lack consistent insight into all programs. The AMRAAM contracts are unusually detailed and allow greater insight than most. Rather than adjust the few programs into which we had sufficient insight and leave the others alone, we chose to leave the costs for all programs as reported.

How This Analysis Aggregated Cost Data

We had several objectives in working with the cost data. Our primary objectives were to

- collect cost data representative of current T&E practices on recent programs
- identify cost trends over time
- identify drivers of system test costs
- develop cost-estimating methodologies useful for cost estimators trying to project resources at the beginning of a development program.

Achieving these objectives was difficult because the two data sets (for aircraft and guided weapons) were diverse, as was the quality of the data. The aircraft data set includes aircraft of various types (fighters, bombers, transport, and trainers) that require different kinds of testing and have different development and unit costs. To allow trend analysis, we included data on a variety of older aircraft.

Similarly, the guided-weapon data set includes weapons of various types and various degrees of new development. The data set includes unpowered weapon kits, such as the Joint Direct Attack Munition (JDAM) and the Wind-Corrected Munitions Dispenser (WCMD), as well as air-to-air and air-to-ground missiles. The scope of development efforts ranges from missile modification programs to full development programs. As with aircraft, weapons of different types and with different amounts of development require different kinds and amounts of testing.

The quality of the data also affected the way we aggregated them and our ability to pursue questions analytically. Contractor costs for aircraft ST&E had the highest level of detail in general, but the WBS elements were not always consistent. Contractor costs for guided-weapon ST&E ranged from great detail for the AMRAAM programs to a single cost number for other programs. Government costs varied similarly.

In light of these realities of the nature of the cost data, we

- identified cost trends over time through discussions with experienced government and contractor test personnel and examined cost trends at the total ST&E level by including legacy programs
- identified drivers of system test cost at a high level, consistent with the quality of the data and diversity of programs, and aggregated data accordingly
- aggregated data so that costs could be explained by variables typically available to estimators at the beginning of a development program.

We grouped aircraft contractor T&E costs into ground test, flight test, other test costs, and a subset of ground-test costs consisting of static and fatigue test costs. These groupings allowed identification of differences in the cost of each group by aircraft type and by characteristics of the test program. When they were available, government T&E costs for aircraft had to be combined into a single figure because they were generally provided to us in an aggregation that did not allow consistent identification of tasks.

Contractor and government costs for guided weapons were each combined into a single figure for analytic purposes. This was done for two reasons. First, in contrast to the case with aircraft, the majority of T&E costs for most guided-weapon programs are for government activities. However, the nature of the T&E effort done by the government varies from program to program, so only a total of both costs consistently represents the entire test effort. Second, aggregation at this level is more consistent with the objective of developing CERs for analysts with limited knowledge of the program at the beginning of a development effort.

After aggregating the cost data as described above, we attempted to generate CERs for each aggregation. CERs were developed for aircraft contractor ST&E costs both in total, as well as ground, flight, and other tests. A separate CER was also developed for static and fatigue testing. Unfortunately, we were unable to obtain reliable government T&E cost data for the legacy aircraft programs that were used to supplement the data collected as part of this study. As a result, there were too few observations of government costs over the various aircraft types to develop CERs for government costs. (The available government cost data appear in the limited-access technical report.)

Because of the diversity of the weapon development programs, we were unable to generate satisfactory CERs for the entire group of weapon programs. However, we were able to develop satisfactory CERs for the guided missiles. Our CER development efforts and results are discussed in Chapter Five.

Estimating Test and Evaluation Costs

The estimating resources in this report are intended to address two situations. The first is when an estimate is needed early in the process of defining a test program, before detailed planning information is available. For example, estimators trying to project the amount of resources for programming or budgeting purposes before SDD may find the methodologies useful. The second is as a cross check to assess the adequacy of resources planned, programmed, or budgeted for a test program by comparison with actual costs on similar programs.

This chapter discusses uses of the data we collected for cost analysis. The following topics are addressed:

- approaches to estimating T&E costs
- data analysis
- CERs.

Approaches to Estimating Test and Evaluation Costs

Because T&E is a complex set of activities tailored to each program, no single estimating method will give the definitive answer. Cost estimators generally use one or more of the following estimating approaches:

- analogy to a similar program
- parametric CERs derived from multiple historical programs
- summations of costs estimated at a detailed level.

The data and methodologies in this monograph support the first two approaches.

For estimating by analogy, the test program descriptions and the limited-access supplement can be used to develop either a cross-check or a primary estimate. If several programs are found to be similar to the one being estimated, the analyst can use the data to develop custom CERs for the case being estimated. For those without access to the supplement, Table 5.1 gives average values for the costs of common T&E elements and other characteristics of potential interest for four representative multiengine fighter programs. Unfortunately there were not enough consistently categorized data to develop similar tables for other types of aircraft or for guided weapons.

For estimating by CERs or factors, we derived quantitative relationships when sufficient data were available and satisfactory relationships could be developed. The aircraft and missile databases contain diverse sets of programs of varying types, cost, and complexity. Obviously, the CERs can be most useful when the estimated program is similar to those in the database.

Estimating by detailed buildup is normally done by test personnel who have knowledge of the test activities to be estimated and the resources that should be required. These estimates are generally developed by the test activity, program office, and system contractor.

In general, we recommend that estimators outside of test organizations use a combination of the analogy and CER approaches to estimate a test program. Because each test program is inherently unique, CERs cannot reflect all the peculiarities of a given test program but can provide an appropriate benchmark for similar programs. Carefully chosen analogies may be the best approach when similar historical programs can be found. Even when there are significant differences, useful comparisons to the historical data can often be made. At a minimum, such comparisons can provide either a cross-check or a starting point for a more in-depth assessment.

Other metrics, such as flight-test hours (Figure 5.1), average flight-test hours per test aircraft month (Figure 5.2), and number of guided-weapon launches per month (Figure 5.3), can also be helpful

Table 5.1
Average Test Costs for Representative Multiengine Fighters
(contractor system T&E FY 2001 $M)

	Average[a]	Standard Deviation	Coefficient of Variation	% of ST&E
Nonrecurring development	2,595.8	494.7	0.19	
Nonrecurring development (less ST&E)	1,815.1	451.2	0.25	
System T&E	795.9	59.0	0.07	100
Wind tunnel test program	37.8	14.3	0.38	5
Static articles and tests	57.6	31.3	0.54	7
Fatigue articles and tests	42.6	6.5	0.15	5
Drop and accelerated loads tests	24.7	1.5	0.06	3
Air vehicle subsystem test program	55.8	14.7	0.26	7
Avionics integration tests	43.4	13.5	0.31	5
Armament and weapon integration tests	7.4	4.0	0.54	1
Mockups	20.3	11.5	0.57	3
Other ground	11.9	8.4	0.71	1
Ground test subtotal[a]	267.4	20.5	0.08	34
Contractor flight tests	367.8	33.4	0.09	46
Other test subtotal	160.8	54.2	0.34	20
Weight empty (lbs.)		28,998	6,203	0.21
Aircraft months		263	28	0.11
Test duration (months)		46	9	0.19
Flight hours		3,771	1,164	0.31
Flight hours per aircraft per month	14.3	3.9	0.28	
Contractor ST&E spent by first flight (%)	44.8	11.2	0.25	

[a]Note that the ground-test subelements do not sum to the ground-test subtotal. In the few cases for which subelements for a particular aircraft had zero values, they were excluded from the calculation of the averages to avoid distorting the averages of the individual subelements.

for attempting to assess the realism of a planned test program based on the experience of previous programs.

In developing parametric CERs, we began with a visual analysis of plots and graphs, then used statistical analysis (linear regression) to develop the CERs. We had three criteria for selecting explanatory variables to predict ST&E costs:

Figure 5.1
Total DT Flight Hours

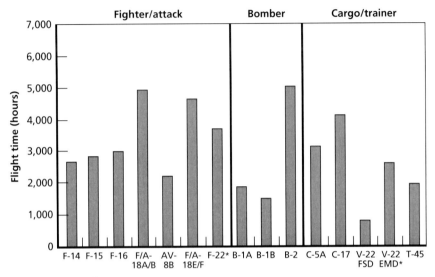

NOTE: * Program in progress.
RAND *MG109-5.1*

Figure 5.2
DT Flight Hours per Aircraft Month

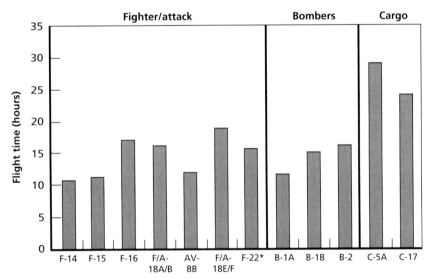

NOTE: * Program in progress.
RAND *MG109-5.2*

Figure 5.3
Guided DT Launches per Month in Order of Contract Award Date

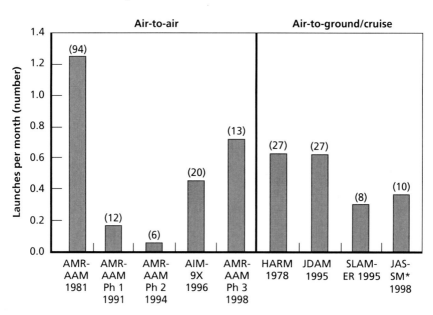

NOTE: * Program in progress.
RAND *MG109-5.3*

- The explanatory variable must have a logical relationship to cost based on the research we had done for the study.
- It must, in general, be available to or determinable by estimators early in a development program.
- It had to be statistically significant in the regression equation.

   In addition to these criteria, we attempted to select variables that could be useful for estimating modification programs and for estimating new development.[1]

   The data we used for CER development included some legacy programs in addition to the more-recent programs collected for this

---

[1] Unfortunately, we had only limited success in obtaining usable T&E cost data on aircraft modification programs (the B-1 Conventional Mission Upgrade Program [CMUP] and the T-45 Cockpit-21 program were exceptions), so we derived the CERs from full development programs.

study. Limiting the data set to recent programs would not have provided sufficient data for meaningful CERs. Expanding the data set to include older programs seemed reasonable because the relationships between our selected cost drivers and the contractor T&E costs of legacy aircraft programs appear to be generally consistent with more recent data. Unfortunately, while government costs were collected for the programs included in this study, they were generally not available for the legacy aircraft programs. As a result, the aircraft CERs include contractor costs only. The cost behavior of weapon programs also appeared roughly consistent over time, so our data set included selected legacy programs.

A Priori Expectations of Aircraft ST&E Relationships

These criteria resulted in a short list of candidate variables for aircraft ST&E costs. Table 5.2 summarizes aircraft test program characteristics. For aircraft ground-test costs, we expected the weight and/or unit cost to be good candidates. We expected weight to be the best predictor of static and fatigue test costs.

These tests require building ground-test articles that are structurally representative of the aircraft, which is itself costly. A cagelike structure is built around the test article to hold the controls and instruments required to conduct the structural tests. Fatigue testing applies representative forces to the airframe structure in repetitive cycles to simulate accelerated structural aging. These tests may last for over a year. The test equipment for a large cargo aircraft is the size of an aircraft hangar and weighs tons. Weight is commonly used as a good predictor of the cost of manufacturing an airframe and is clearly related to the test equipment used in this subset of tests.

Other ground-test costs include subsystem, avionics, and propulsion integration tests. We expected the cost of the airframe and these subsystems to drive test costs. Because the cost of the airframe,

Table 5.2
Summary of Aircraft Test Program Characteristics

Aircraft	Months	Duration[a]	Flight Hours (DT)	Average Flight Hours per Aircraft Month	Empty Weight	First Flight Year
F-14A	250	45	2,685	10.7	36,825	1970
F-15A/B	253	59	2,856	11.3	26,250	1972
F-16A/B	156	25	2,581	16.5	13,371	1976
F/A-18A/B	305	40	4,922	16.2	22,351	1978
AV-8B	147	37	2,038	13.9	12,931	1981
F/A-18E/F	244	41	4,620	18.9	30,564	1995
F-22	N/A	N/A	N/A	N/A	31,670	1997
B-1A+B	264	131	3,425	13.0	181,400	1974/ 1984
B-2	310	95	5,032	16.2	153,700	1989
C-5A	108	61	3,145	29.2	320,085	1968
C-17	169	39	4,104	24.3	269,696	1991
V-22 FSD	N/A	43	763.6	N/A	31,886	1989
V-22 EMD	N/A	N/A	N/A	N/A	33,140	1997
T-45	N/A	64	1,932	N/A	9,394	1988
B-1 CMUP	32	11	712	22.3	181,400	1997

[a]Months from first flight to end of DT.

subsystems, and avionics may not be available separately at the beginning of a program and because these aircraft elements constitute much of the cost of an aircraft, the unit cost for the air vehicle (excluding propulsion cost) is a reasonable substitute.[2]

We expected aircraft flight-test costs to have both fixed and variable components. A flight-test program requires a certain amount of staff and equipment infrastructure. Once the infrastructure is in place, its costs are incurred regardless of whether or not the aircraft are flown. These costs are relatively fixed and cannot be readily turned on and off in response to short-term variations in test activity. In addition to the fixed costs, there are variable costs for personnel, test and support aircraft, and such expendables as fuel and parts. Thus we expected that the best explanation of the costs of flight test-

___
[2] We calculated the theoretical first unit (T_1) cost using an 80-percent unit theory cost-improvement curve. Note that only data from FSD/EMD lots were used to develop T_1 costs.

ing would be some measure of the fixed nature of the flight-test effort, such as length of DT, and some measure of the amount of flying, such as number of flights or flight hours, or a variable that captures both fixed and variable aspects, such as test aircraft months.

It was more difficult to find an intuitive cost driver for aircraft "other test" costs. One reason for the difficulty is that this group of costs includes rather amorphous WBS elements found in most aircraft development programs, such as T&E support and other system T&E. These are various T&E costs that cannot be clearly assigned to either ground or flight-test activities. In attempting to explain this miscellaneous group of costs that are related to both ground and flight-test events, we expected them to be a function of both ground and flight-test activity or of nonrecurring development cost.

In recognition of the argument made by some test managers that test programs should be a set of integrated and synergistic efforts that should not be arbitrarily separated or considered to be isolated, independent activities, we also developed CERs for ST&E at the total contractor cost level. We again selected explanatory variables, such as weight or unit cost, that seem to drive ground tests and variables, such as flight hours, aircraft months, or DT duration, that we expect drive the fixed and variable flight-test costs. We also examined nonrecurring development cost as a measure of the overall complexity of the development effort and, by implication, of the test program.

Results of Aircraft ST&E Analysis

Using these variables that met our first two criteria of having a logical relationship to test costs and being generally available to cost estimators, we obtained reasonable statistical relationships for total ST&E, ground test, static and fatigue test, flight test, and other test costs.[3] Definitions and abbreviations of the variables are shown in Table 5.3.

---

[3] The values of the input parameters for many of the CERs may change as the development program proceeds. When estimated parameter values are used, they should be varied over an appropriate range to examine sensitivity in the predicted values.

The correlation matrix for the variables is shown at Table 5.4 and uses the same abbreviations.

The CERs presented below as "preferred" best met our selection criteria. In addition, we developed alternative CERs using different independent variables, formulations, or a subset of the available data for use in estimating situations when these constructions may be more appropriate.[4] All the CERs have F-statistics for the equation and t-statistics for the independent variable that are significant at the 95-percent confidence level,[5] unless otherwise noted. They express contractor costs in constant FY 2001 million dollars.

Table 5.3
Definitions of Aircraft Variables

Variable	Variable Definition	Units
NR DEV	(Nonrecurring development less test) Contractor's total cost for the nonrecurring portion of the aircraft development program, excluding the cost of ST&E	FY 2001 $M
STE	(Contractor ST&E) Contractor's total cost for ST&E, excluding the government's cost for DT or OT	FY 2001 $M
GND[a]	(Contractor ground test) Costs for contractor ground testing (wind tunnel, static, fatigue, drop, subsystem, propulsion integration, avionics integration, weapon integration, mockups, and similar tests)	FY 2001 $M
FLT[a]	(Contractor flight test) Costs for contractor flight testing (includes only events explicitly labeled as flight tests in the contractor's cost reports)	FY 2001 $M
GND + FLT[a]	(Contractor ground and flight test) Contractor's total cost for ground and flight testing, as defined above	FY 2001 $M
OTHER[a]	(Contractor other test) Contractor's T&E cost for tests not included in ground or flight testing (e.g., T&E support and test requirements)	FY 2001 $M

---

[4] As previously noted, the lack of government T&E cost data for legacy programs precluded development of CERs for government costs.

[5] *F-* and *t-statistics* are measures of the significance of the coefficients of the entire equation and the individual coefficients, respectively. *Adjusted r^2* is a measure of the variance "explained" by the selected equation, adjusted for the degrees of freedom. The coefficient of variation is the estimated standard error of the regression equation divided by the mean of the dependent variable.

Table 5.3—continued

Variable	Variable Definition	Units
ST/F[a]	(Static and fatigue) Contractor's T&E costs for static and fatigue testing; includes the cost of building the test articles and conducting the tests	FY 2001 $M
T_1	(T_1 cost) Theoretical first unit (T_1) cost for the flight-test units built during the development program[b]	FY 2001 $M
WE[c]	(Weight empty) Total weight of the aircraft structure and its subsystems, avionics, and engine	Pounds
AC Mo	(Aircraft months) Total time each flight-test aircraft is available for flight testing during DT	Months
EMD DUR	(EMD duration) Elapsed time from development contract award to end of DT	Months
DUR	(Flight test duration) Elapsed time from first flight to end of DT	Months
FLT HRS	(Flight hours) Accumulated flying time during DT for all flight-test aircraft	Hours
F/A	(Fighter/attack) Dummy variable used in regression analysis to distinguish among different classes, in this case, between fighter or attack aircraft and other aircraft	1 = fighter/attack aircraft 0 = non–fighter/attack aircraft
CGO	(Cargo) Dummy variable used in regression analysis to distinguish among different classes, in this case, between cargo and other aircraft	1 = cargo aircraft 0 = non–cargo aircraft

[a]Subtotals.
[b]Calculated by dividing the total recurring air vehicle cost in development by the number of FSD and EMD units, assuming an 80-percent cost improvement curve, using the curve to calculate an algebraic lot midpoint, and backing up the curve from that point to the T_1 cost:

$$T_1 \text{ cost} = \frac{\text{Average unit cost}}{(\text{Lot midpoint})^{-0.322}}$$

Lot midpoint formulae can be found in various cost estimating or economics texts.
[c]According to MIL-STD-1374 (DoD, 1977), weight empty is more inclusive than either structure or airframe unit weight.

The data set for the CERs contains the following aircraft development programs: AV-8B, B-1A and B-1B (combined), B-2, C-5A, C-17, F-14, F-15, F-16, F/A-18A/B, F/A-18E/F, and T-45. The

AV-8B and F-16 programs include cost and programmatic information from their prototype programs. We excluded the F-22 and V-22 programs from the data set used for generating CERs, even though we collected information on the programs, because they are in progress and costs are not complete. We omitted the B-1 program from the data set for generating some flight test and ST&E CERs because the B-1A's flight-test program was unusually long, resulting in atypical costs. Although the B-2's values were considerably larger than those for the other programs, our analysis included this aircraft because it provides the only example of both a stealthy aircraft and a continuous development effort for a modern bomber. As summarized in Table 5.5, CERs for five categories of test costs—total contractor ST&E, ground test, static and fatigue tests (which are a subset of ground test), flight test, and other test—are presented below.

Following that table is a series of subsections that present the CERs we developed for each category with a short discussion of each.

Table 5.4
Aircraft Variables Correlation Matrix (11 Programs)

	NR DEV	STE	GND	FLT	GND + FLT	OTHER	ST/F	T_1	WE	AC Mo	PGM DUR	DUR	FLT HRS	F/A	CGO
NR DEV	100														
STE	93	100													
GND	92	97	100												
FLT	63	80	64	100											
GND + FLT	91	100	97	81	100										
OTHER	95	99	95	77	97	100									
ST/F	61	63	91	14	67	50	100								
T_1	96	95	98	58	94	95	94	100							
WE	40	42	58	8	47	33	99	51	100						
AC Mo	48	53	36	75	52	54	-35	36	-35	100					
EMD DUR	74	85	78	81	86	81	52	73	47	42	100				
DUR	54	76	67	82	77	71	40	58	35	42	94	100			
FLT HRS	63	58	58	46	59	55	28	58	28	61	35	16	100		
F/A	-51	-52	-61	-24	-54	-47	-71	-58	-74	10	-73	-63	-11	100	
CGO	-11	-22	-3	-51	-19	-26	54	-3	58	-62	1	-7	-19	-67	100

Total Contractor ST&E

Figure 5.4 compares the actual and predicted costs of the preferred CER. Table 5.6 is the preferred CER, which excludes the B-1A/B program on the assumption that its cancellation and restart intro-

Table 5.5
Summary of T&E Estimating Resources

	CER (Preferred)	CER (Alternative)	Database
Contractor ST&E	●	●	●
Ground tests	●	●	●
Static and fatigue test	●		●
Other ground test			●
Flight tests	●	●	●
Other tests	●		●
Government DT&E			●
Government OT&E			●

Figure 5.4
Total Contractor ST&E Cost CER

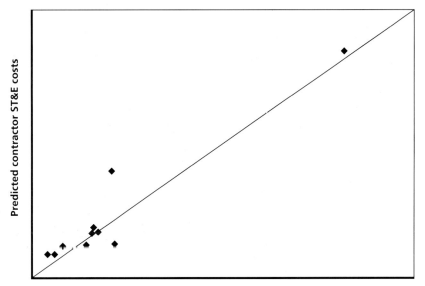

duced atypical inefficiencies. Tables 5.7 and 5.8 present alternative CERs. Table 5.7 does include the B-1. The preferred and alternative CERs suffer from the limitation of requiring an estimate of non-recuring development cost as an input.

Table 5.6
Total ST&E CER—Preferred

CER	STE = 215.9 + 0.2757NR DEV
Standard error	357.11
F-statistic	78.91
t-statistic on independent variable	8.89
Coefficient of variation	34.62%
Adjusted r^2	89.64%
Number of observations	10

NOTE: Omits B-1.

Table 5.7
Total ST&E CER—Alternative 1

CER	STE = 1.828(NR DEV)$^{0.806}$
Standard error	463.55
F-statistic	42.72
t-statistic on independent variable	6.54
Coefficient of variation	39.59%
Adjusted r^2 in unit space	83.75%
Number of observations	11

NOTE: Includes B-1.

Table 5.8
Total ST&E CER—Alternative 2

CER	STE = 2.509(NR DEV)$^{0.7586}$
Standard error	428.78
F-statistic	35.36
t-statistic on independent variable	5.95
Coefficient of variation	41.57%
Adjusted r^2 in unit space	85.07%
Number of observations	10

NOTE: Omits B-1.

Contractor Ground Testing

Figure 5.5 compares the actual and predicted contractor ground test costs of the preferred CER, shown in Table 5.9. Table 5.10 presents an alternative CER using nonrecurring development cost and empty weight.

The T_1 cost for the FSD/EMD units proved to be the best predictor of ground-test costs. Although weight was a good predictor for static and fatigue tests, the recurring unit cost of the aircraft is a better predictor when other ground-test costs, such as subsystem and avionics integration, are included. The CER has the undesirable feature of requiring an estimated first unit cost to generate the estimate of ground-test costs.

Figure 5.5
Contractor Ground-Test CER

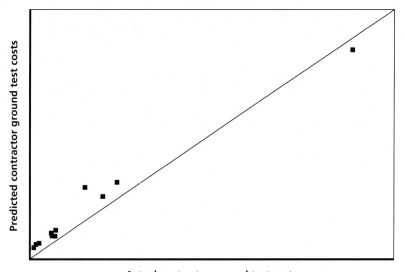

RAND *MG109-5.5*

Table 5.9
Ground-Test CER—Preferred

CER	$GND = 94.84 + 1.54T_1$
Standard error	108.27
F-statistic	279.10096
t-statistic on independ- ent variable	16.706
Coefficient of variation based on standard error (standard error/avg act)	20.86%
Adjusted r^2	96.53%
Number of observations	11

Table 5.10
Ground-Test CER—Alternative

CER	$GND = (-13.32) + 0.1299NR\ DEV + 0.001278WE$
Standard error	197.69
F-statistic	39.21
t-statistic on first inde- pendent variable	7.02
t-statistic on second independent variable	2.13
Coefficient of variation	38.10%
Adjusted r^2	88.43%
Number of observations	11

NOTE: Empty weight is significant at 93 percent.

Static and Fatigue Testing

Static and fatigue test costs are a significant portion of ground-test costs. Because these costs are classified and reported relatively consistently, we were able to develop a single satisfactory CER (Figure 5.6 and Table 5.11).

Empty weight is the independent variable. Our data source for the B-2 did not identify the cost of static and fatigue tests, so we could not include it in the data set.

Figure 5.6
Static and Fatigue Test Costs

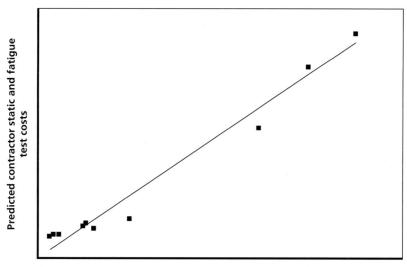

Actual contractor static and fatigue test costs

RAND *MG109-5.6*

Table 5.11
Static and Fatigue Test CER—Preferred

CER	ST/F = 37.21 + 0.001573WE
Standard error	34.35
F-statistic	263.68
t-statistic on independent variable	16.24
Coefficient of variation based on standard error (standard error/avg act)	18.94%
Adjusted r^2	96.69%
Number of observations	10

NOTE: Omits B-2 because of the lack of data on static and fatigue costs.

Flight Testing

The CER in Figure 5.7 and Table 5.12 provided the best predictive value for contractor flight-test costs, capturing both the fixed (duration) and variable (aircraft months) aspects of flight testing. Aircraft months are defined as the total number of months particular aircraft are assigned to the development flight-test program. Duration is defined as the number of months from first flight to end of DT. The above relationship has the undesirable feature of a large negative intercept, which makes it especially important to use the CER within the range of the historical data and to check the results against selected analogies.

Alternative CERs are provided in Tables 5.13 and 5.14. The second alternative CER (Table 5.14), which excludes the B-1A/B, has a similar coefficient of variation and eliminates the negative intercept. (Dropping the B-1 does not significantly change the coefficient of variation in the linear formulation.)

Figure 5.7
Contractor Flight-Test CER

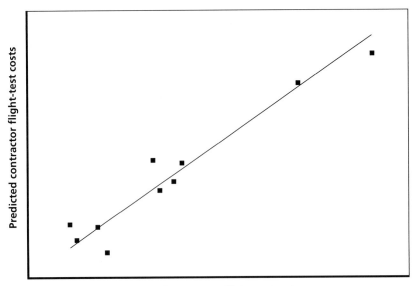

Actual contractor flight-test costs

Table 5.12
Flight-Test CER—Preferred

CER	FLT = (−311.7) + 1.736AC Mo + 5.268DUR
Standard error	75.32
F-statistic	48.51
t-statistic on first independent variable	4.41
t-statistic on second independent variable	6.15
Coefficient of variation	20.20%
Adjusted r^2	91.35%
Number of observations	10

NOTE: Aircraft months were not available for the T-45.

Table 5.13
Flight-Test CER—Alternative 1

CER	FLT = $0.01887(AC\ Mo)^{1.454}(DUR)^{0.4879}$
Standard error	101.90
F-statistic	33.76
t-statistic on first independent variable	6.05
t-statistic on second independent variable	2.78
Coefficient of variation	27.33%
Adjusted r^2 in unit space	84.16%
Number of observations	10

NOTE: T-45 omitted due to lack of data on aircraft months. Duration is significant at 70 percent.

Table 5.14
Flight-Test CER—Alternative 2

CER	FLT = $0.04654(AC\ Mo)^{1.475}(DUR)^{0.2137}$
Standard error	68.82
F-statistic	39.92
t-statistic on first independent variable	7.68
t-statistic on second independent variable	1.15
Coefficient of variation	21.96%
Adjusted r^2 in unit space	86.18%
Number of observations	9

NOTE: T-45 omitted due to lack of data on aircraft months. B-1 omitted because of stretch out. Duration is significant at 81 percent.

Another caveat is that these CERs estimate contractor flight-test costs only. There is some evidence that the amount of government cost and effort on flight-test programs is increasing over time. In fact, government costs have been larger than contractor flight-test costs on the two most recent fighter flight-test programs, the F/A-18E/F and projections for the F-22.

Figure 5.8 shows the percentage of the flight-test programs that government activities perform. The contractor and government efforts were largely separate for the F-14 (first flight 1971) and the F/A-18A/B (first flight 1978).[6] The F/A-18E/F and F-22 programs are from the 1990s and early 2000s, respectively. By this time, test

Figure 5.8
Percentage of Government Costs in Flight Test Have Been Increasing

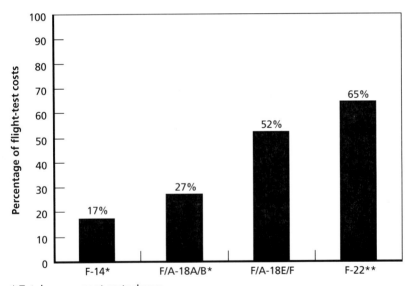

* Total government costs shown.
** In progress.
RAND MG109-5.8

---

[6] In these two cases, the government costs shown represent total program office spending on the program during the flight-test period and are thus an upper bound on what the program office could have spent on testing alone.

programs had become integrated contractor and government efforts, and the programs were paying for most of the variable costs of testing. We did not have government cost data on enough programs to generate a CER that included government costs, but analysts with access to the proprietary supplement can use the CERs for contractor costs in conjunction with the government costs on individual programs along with information on the program being estimated to estimate the total cost of flight test.

Other Contractor Test Costs
Figure 5.9 compares actual and predicted costs, and Table 5.15 presents the CER we developed for other test costs.

This category includes a variety of efforts that are not clearly associated with either flight or ground tests, such as test support, planning, and test requirements. The nature of the activities included in this subtotal makes it logically related to the scope and complexity

Figure 5.9
CER for Other Contractor Test Costs

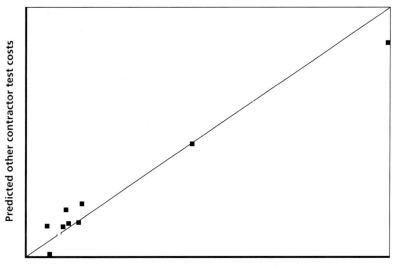

RAND *MG109-5.9*

Table 5.15
Other Contractor Test CER—Preferred

CER	Contractor Other Test = (–134.5) + 0.5041(Estimated Contractor Ground + Flight Test)
Standard error	80.82
F-statistic	225.9
t-statistic on independent variable	15.03
Coefficient of variation	27.13%
Adjusted r^2	95.74%
Number of observations	11

of the rest of the test program. The CER estimates other test costs as a function of the estimated costs of ground and flight test.

A Priori Expectations for Guided-Weapon ST&E Relationships

Our analysis of guided-weapon ST&E costs was more constrained than was our analysis of aircraft test costs primarily because of the variety of classes of systems in our data set. Table 5.16 summarizes the characteristics of the data set. The table shows that the data set of weapon programs includes weapons powered for medium range (air-to-air or air-to-ground missiles) or long-range (cruise missiles) flight, or unpowered; new development programs or modifications of existing weapons; and test programs of varying length and scope.

We identified cost drivers for weapon ST&E from our discussions with personnel in program offices and test organizations and from our analysis of the data collected. Our identification of cost drivers guided the development and selection of CERs, although we were often unable to find a satisfactory quantitative expression of cost relationships because of the familiar (for cost estimators) problems of too many variables for too few observations.

The first cost driver considered for missiles and munitions was the type of weapon. This set of cost drivers divided our data set into three weapon classes. We expected and observed that unpowered

Table 5.16
Summary Characteristics of Missiles and Guided Munitions

	Type of Weapon	Type of Devel. Program	DT Guided or Live Launches	DT Months	FSD/EMD Start
Phoenix	Air-to-air missile	New	64	43	December 1962
AMRAAM	Air-to-air missile	New	86	44	December 1981
AMRAAM Ph 1	Air-to-air missile	Mod	12	24	March 1991
AMRAAM Ph 2	Air-to-air missile	Mod	13	89	June 1994
AIM-9X	Air-to-air missile	Mod	20	33	December 1996
AMRAAM Ph 3	Air-to-air missile	Mod	8	18	December 1998
HARM	Air-to-ground missile	New	23	18	August 1978
IIR Maverick	Air-to-ground missile	Mod	52	21	October 1978
Harpoon	Cruise missile	New	33	15	June 1971
SLAM-ER	Cruise missile	Mod	8	15	March 1995
JASSM[a]	Cruise missile	New	10	19	November 1998
SFW	Unpowered munition	New	25	39	November 1985
JSOW Baseline	Unpowered munition	New	24	N/A	June 1992
JDAM	Unpowered munition	New	170	N/A	October 1995
WCMD	Unpowered munition	New	61	20	January 1997

[a]In progress; not used for CER development.

munitions, such as JDAM, JSOW, and WCMD, were the least expensive to test as a class of weapons. The complex sensor-fuzed weapon (SFW) was relatively expensive for this group. We expected and observed that cruise missiles were more expensive to test than shorter-range missiles on a cost per shot basis.

The second cost driver was whether the program was a new development or a modification of an existing system. Logically, we expected new development programs to be more expensive to test. Among the air-to-air missiles, for example, the AMRAAM pre-planned product improvement (P3I) programs were less expensive to test than the more-ambitious AIM-9X modification program or the original AMRAAM FSD program. This second cost driver further subdivided our data set because some programs within the air-to-air and cruise missile types were new developments and some were modifications.

A third cost driver is the complexity of the total system or missile component that is being developed and tested. Unlike the first two cost drivers, complexity as it relates to testing is not easy to define or to measure consistently. Indicators of complexity include the number and types of sensors and operational modes. Among the unguided munitions, SFW is complex because of its self-contained sensors and the multiple stages the weapon goes through from launch until it fires. It had the largest test costs in total and on a cost-per-shot basis among its class. Among the air-to-air missile modification programs, the AIM-9X made the most extensive changes, with newly developed guidance and tail control systems. It had the highest test costs among modification programs. Typical parameters for measuring the complexity of the system, such as operational modes, weight, and density, were not very useful in characterizing T&E for modification programs, which is more a function of the nature and extent of the modifications performed. To attempt to capture both the degree of modification and system complexity, we used contractor development cost less T&E.

Once weapon programs were categorized by type, new development or modification, and complexity, we expected that the number of flight tests would drive ST&E costs. Flights with launches are more costly than captive-carry flights, and guided launches more costly than separation launches.[7] For example, we expected to find a relationship between ST&E costs and the number of captive-carry flights and live shots for air-to-air missiles. These programs typically

---

[7] It should be noted that certain complex captive-carry flight tests can be nearly as expensive as guided launches.

require a great deal of captive-flight testing to assess the fidelity of simulation models, provide sensor data for the models and software development, test missile aerodynamic and integration issues with the platform, etc. We expected that the relationship between flight tests and test costs for the unpowered weapons that are also part of the data set would be weaker. In general, we expected the number of launches to be a cost driver for each type of weapon. We also expected that the cost per shot would be very different for an unpowered bomb kit with limited range, such as JDAM, and an expensive, long range, cruise missile, such as JASSM or the Standoff Land-Attack Missile–Expanded Response (SLAM-ER), even ignoring the cost of the test article.

Results of Weapon ST&E Analysis

We began our efforts to develop CERs from the entire data set of weapon programs. However, the cost behavior of the unpowered weapons (JDAM, JSOW, WCMD, and SFW) could not be satisfactorily modeled with the available data. Dropping the unpowered weapons from the data set resulted in a satisfactory CER for total (contractor and government) guided-weapon test costs for missiles. Table 5.17 defines the variables for missiles and contains the abbreviations used in Table 5.18. Table 5.18 is a correlation matrix.

The preferred CER for missiles (Figure 5.10 and Table 5.19) estimates total DT costs as a function of contractor nontest development costs and number of launches. This form accommodated both new development and modification programs and reflected the variable costs in the number of launches.

We developed an alternative CER (Table 5.20) using number of launches and DT duration, dropping AMRAAM Phase 2 because of its unusually long duration. The other candidate independent variables were not as robust across the range of systems.

Table 5.17
Variable Definitions for Missiles

Variable Name	Variable Definition	Units
CTR DEV	(Contractor development less test) Contractor cost for the missile development program[a] minus the cost of ST&E	FY 2001 $M
CTR TEST	(Contractor test) Contractor test costs in; does not include government costs of DT or OT	FY 2001 $M
GOVT DT	(Government DT) Government costs for DT; does not include the cost of OT	FY 2001 $M
STE	(ST&E) Total contractor and government test costs, excluding OT	FY 2001 $M
LAUNCH	(Launches) Guided-weapon launches in DT, not including separation test launches or captive carry flights	Number
DUR	(Duration) Elapsed time from first flight or launch to end of DT	Months
MOD	(Modification) Dummy variable for program type	1 = modification program 0 = new development
AA	(Air-to-air) Dummy variable for missile type	1 = air-to-air missile 0 = other missile

[a]Many cost reports used for weapons in this study did not differentiate between non-recurring and recurring costs.

Table 5.18
Missile Variables Correlation Matrix (10 Programs)

	CTR DEV	CTR TEST	GOVT DT	STE	LAUNCH	DUR	MOD	AA
CTR DEV	100							
CTR TEST	64	100						
GOVT DT	84	31	100					
STE	92	59	95	100				
LAUNCH	77	50	90	93	100			
DUR	14	−17	27	17	13	100		
MOD	−79	−44	−67	−71	−63	8	100	
AA	23	−11	29	21	9	56	17	100

Figure 5.10
Missile Total ST&E CER (Excludes Guided Munitions)

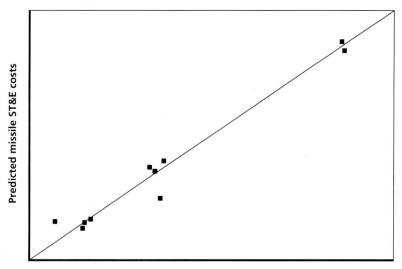

Actual missile ST&E costs

Table 5.19
Missile Total ST&E CER—Preferred

CER	STE = 15.34 + 0.07078CTR DEV + 1.401LAUNCH
Standard error	14.35
F-statistic	103.82
t-statistic on first inde- pendent variable	4.77
t-statistic on second independent variable	5.02
Coefficient of variation	15.60%
Adjusted r^2	95.81%
Number of observations	10

NOTE: Includes missile programs only. The equations and the independent variables are significant at the 98 percent level of confidence or higher except where noted.

Table 5.20
Missile Total ST&E CER—Alternative

CER	STE = (−18.35) + 1.713LAUNCH + 2.244DUR
Standard error	24.99
F-statistic	30.21
t-statistic on first independent variable	3.56
t-statistic on second independent variable	1.95
Coefficient of variation	25.64%
Adjusted r^2	87.96%
Number of observations	9

NOTE: Omits AMRAAM Phase 2; duration is an outlier. The duration variable is significant at 90 percent.

Developing a T&E Cost Estimate

As discussed above, the estimating sections of this report were written primarily for cost analysts who are tasked either with developing estimates of test programs before detailed test planning data are available or with developing independent estimates and assessments of program office T&E estimates.

The first step in developing an estimate for a test program is to determine its likely scope. Table 5.21 provides some indicators that, in general terms, distinguish relatively complex test programs from simpler efforts.

The primary missions of an aircraft or guided weapon are general indicators of its complexity. Fighter aircraft and air-to-air or cruise missiles are more difficult to develop and test than trainers or guided munitions. However, within broad missions, there can be varying levels of expected performance, which will affect the amount of development and therefore the level of testing involved. The testing complexity for a fighter aircraft—achieving maximum levels of speed, maneuverability, integration of onboard and offboard systems, and signature reduction—can be significantly greater than for a program whose performance requirements are less challenging.

In addition to mission- and performance-driven complexity, the amount of new development in mission critical systems and subsystems directly affects test programs. The programs we examined for this study provide examples of high new-development content (F-22, B-2, V-22, AMRAAM, SFW) and of more-evolutionary systems (F-18E/F, B-1B CMUP, SLAM-ER, AMRAAM P3I). The system's Cost Analysis Requirements Description (CARD) or other program documentation usually indicates which subsystems or key components have been proven in other applications.

The planned activities and their schedule constitute a third class of indicators of test program content. A general outline of the required test program should be available early in the acquisition process. Such test requirements as static and fatigue, full-scale live-fire, extensive avionics and software, interoperability, LO, or a full range of sensor and seeker testing are key indicators of test scope and risk. For aircraft, developing a new engine is a major program in itself, with corresponding test requirements.[8] The planned duration of the overall test program is another indication of both content and risk.

Table 5.21
Indicators of Test Program Scope

Indicator	More-Complex Examples	Less-Complex Examples
Mission	Air superiority	Air transport
	Air-to-air	Munition kit
System	Maximum performance	Limited capability
	Extensive integration	Stand-alone
	LO	Conventional signature
	Extensive functionality	Single purpose
	New technology	Proven design
Design	New design	Limited enhancement
Test requirements	Unusual test environments	Typical environments
	Compressed schedule	Available slack
	Multiservice	Single service
	Unusual security	Typical security

Test planning is a continual balance of likely cost and risk. Many aspects of testing have costs that are more a function of time

---

[8] This monograph addresses only the integration testing of the engine and aircraft.

than of activity. Test schedules may be compressed in an effort to reduce both testing and development costs. If this incentivizes higher productivity and innovation on the part of government and contractor testers, it can be an effective approach. However, arbitrary schedule compression, especially when there are delays outside the control of the test community, such as late delivery of testable hardware or delays and retesting for correction of defects, commonly results in slipped schedules and increased costs.

After determining the likely scope of the test program, the next step is to compare the planned program with similar historical cases. As discussed throughout this monograph, T&E processes and approaches continue to evolve. However the realism of program plans that project large T&E savings must be carefully evaluated in view of historical experience. Appendix A briefly describes the technical, programmatic and T&E aspects of eight aircraft and eight guided-weapon programs. Information on some older programs may be found in several studies listed in the References. We have attempted to provide enough information to allow the analyst to draw meaningful comparisons of test program scope, schedule, and cost. The corresponding cost data are contained in a limited-access supplement to this report. Using these resources, programs of similar content can be identified.

As a first step, an estimate of total contractor T&E cost for aircraft programs can be made using the parametric CERs provided above. If sufficient test program information is available, this estimate can be refined considerably by using the individual CERs for ground, flight and other test costs. Estimates of government costs can be developed by analogy with similar programs from the data supplement. For missile programs, the preferred and alternative CERs can be used to estimate both the contractor and government portions of the DT program.

This estimate should then be compared to the actual costs and durations of similar programs in the database. In the best case, several programs may be similar in key aspects to the program being estimated. Rudimentary data analysis will provide averages and ranges for total test costs and durations, highlighting outliers. When more-detailed data are given, major test activities can also be compared directly. When adequate data are available, regression analysis can be used to develop tailored CERs at the total level or for selected portions of the test program. If sufficient comparable data are not available to derive meaningful statistical relationships, it may be useful to use averages and ranges for several analogous programs or values from a single comparable program, adjusted for differences in the program being estimated.

As always when dealing with small data sets, cross checks should be used to confirm estimates. Alternative metrics that can be useful to develop cross checks include the following:

- cost per flight hour (aircraft)
- cost per shot (guided weapons)
- cost per test month.

CHAPTER SIX

Conclusions and Recommendations

The following conclusions can be drawn from the data we collected and our interviews with government and contractor test personnel:

- The overall cost of T&E has been a relatively constant proportion of aircraft and guided-weapon system development costs since the early 1970s. Despite increasing use of M&S, improvements in instrumentation and test processes, reduction of dedicated government testing, and various acquisition streamlining initiatives, T&E costs have remained relatively consistent. Although various explanations of this situation are possible, the dominant factors are probably the increasing complexity of the systems being tested and greater test program content.
- M&S is now integral to most test programs. In fact, in several cases, M&S capabilities were hard pressed to meet the program's T&E requirements. In many programs, the analytical tools were not always adequate to confidently waive live testing, but in all cases, M&S was judged to be a good investment that at least reduced the risk, and often the duration, of live tests. In addition to direct benefits for T&E, robust M&S has other benefits for

  - evaluating design excursions during development
  - developing tactics
  - training operators
  - evaluating future system enhancements.

- Although there is some disagreement about the appropriate level of testing in specific circumstances (e.g., live-fire testing, testing for statistically rare events), we found little controversy in general over the scope of testing. Several sources, however, expressed the opinion that thoughtful reevaluation of test procedures could improve the pace and efficiency of the typical test program.
- There was general agreement that integrated contractor-government test teams were a positive force for optimizing testing. Similarly, combined development and operational test teams were judged to have been valuable because they avoid redundant testing and identify operational effectiveness and suitability issues for early resolution. Some program personnel expressed a desire for even more intensive "early involvement" from the operational test community. The primary constraint appears to be the limited staffing of the services' operational test organizations.
- It is too early to assess the outcome of some recent innovative test management approaches that involve giving the contractor broad latitude in developing and executing the DT program. Another innovative approach—relying on non-DoD tests and certifications of nondevelopmental aircraft for DoD applications—has not been as successful as hoped. In the two cases we examined (both trainer aircraft), the DoD requirements were different enough from those of previous customers to require significant modification and testing. In both cases, the problems were more the result of underestimating the differences in requirements and the scope of required modifications than of quality problems with previous testing and certification processes.
- Data on costs incurred by government organizations are much more difficult to collect and document than are the corresponding contractor cost data. This did not initially seem to be a serious limitation because we assumed that acquisition reform would lead to decreasing government costs as contractors took on a greater share of the effort. For programs where we were able

to obtain government costs, this was not generally the case. In many instances, contractors still rely on government test facilities and functional expertise, particularly for high-cost, low-utilization test capabilities. Government personnel normally participate actively in the integrated test teams. Even in programs that do not constrain the contractor's choice of test facilities, the government facilities can find themselves acting as test subcontractors to the system prime contractor. Of course, most open-air testing continues to take place on DoD ranges. Our recommendation is that government cost data be consistently accumulated and reported, just as contractor data is today. This would ensure that the program's total financial picture would be available for management in the present and analysis in the future. This would help government test facilities better evaluate the cost and schedule implications of their processes, allowing them to better assess the contribution of all their activities and to focus investment and management attention on those deemed most critical to their customer base.

Aircraft Program Descriptions

B-1B Conventional Mission Upgrade Program Block D

Mission
The B-1B is a long-range supersonic bomber, originally designed to penetrate sophisticated air defenses.[1] The CMUP enhances the B-1B's capability to deliver modern conventional weapons. Modifications are being incorporated in four combined hardware and software block upgrades. This description summarizes only the CMUP Block D upgrade test program.

System Description
The Block D upgrade to the B-1B consists of the following hardware and software modifications:

- Global Positioning System (GPS) capability for navigation, offensive avionics, and weapon delivery
- jam-resistant very high and ultrahigh frequency radios
- MIL-STD-1760 interface to the Multipurpose Rotary Launcher to accommodate advanced conventional munitions
- JDAM integration
- offensive system and ancillary software sustainment upgrades

---

[1] Information on the B-1B came from SPO input, Air Force Flight Test Center (2000) for DT/IOT&E, and AFOTEC (1998) for dedicated OT&E.

- defensive system software improvements.

Programmatics

Boeing North American, the system prime contractor, was awarded the contract to integrate the B-1 CMUP modifications. Programmatic milestones included the following:

- EMD contract award: March 1995 (Contract F33657-94-C-0001)
- critical design review (CDR): May 1996
- production approval: July 1997
- production contract award: July 1997 (Contract F33657-97-C-0004).

Test Program

The combined DT&E/IOT&E focused on verifying the GPS incorporation, sustainment software upgrades, reincorporation of ground moving target indication and tracking, defensive system software upgrades, and JDAM integration.

Initial avionics testing was conducted at the contractor's avionics and system integration laboratories, the Avionics Integrated Support Facility at Tinker AFB, and the Integrated Facility for Avionics System Test at Edwards AFB. The avionics software modifications were accomplished in 18 months because software anomalies were identified during ground and flight tests.

Initial JDAM testing took place at Arnold Engineering Development Center, followed by ground testing; captive-carriage, safe-separation, environmental conditions, safety, and electromagnetic interference (EMI) and electromagnetic compatibility (EMC) testing; and actual flight tests at Edwards AFB. Approximately six JDAM separation test vehicles and 23 guided test vehicles were dropped. In addition to testing the performance of the Block D changes, the test program also verified correction of previously identified system-level deficiencies. Full-up live-fire testing of the B-1B was waived, but component-level vulnerability testing was performed. There were 5,411 hours of major ground testing.

Table A.1 summarizes the aircraft sorties made for the various types of testing. Table A.2 provides data for the participating aircraft. The three aircraft put in a total of 34 aircraft months of service.

AFOTEC conducted an operational assessment, combined DT&E/IOT&E, and a one-month dedicated IOT&E. Approximately nine sorties each were used for combined testing and dedicated IOT&E.

EMD began in January 1995, with the first test flight in March 1997. DT ended in September 1998; two aircraft made 97 flights, flying for 660 hours. On average, the flight-test program achieved 20.6 flight hours per aircraft month, using engine-running crew changes to maximize aircraft utilization.

Dedicated OT&E began in August 1998 and ended in September 1998. The three aircraft mentioned in Table A.2 provided six air-

Table A.1
B-1B Flight Testing

Event	Test Aircraft (no.)	Sorties (no.)	Flight Time (hrs.)
Total EMD flight test August 1997–September 1998	3	106	712
DT&E/IOT&E August 1997–July 1998	2	97	660
Dedicated IOT&E August 1998–September 1998	2	9	52

Table A.2
The B-1B Test Aircraft

Test Aircraft	Entered Testing	Exited Testing	Aircraft Months
85-0068	March 1997	September 1998	18
04-0040	July 1997	September 1998	14
85-0082[a]	August 1998	September 1998	2

[a]Used for dedicated OT&E.

craft months and flew a total of nine flights and 53 flight hours. Aircraft 85-0082 was the Block D Kit Proof aircraft.

B-2A Spirit

Mission

The B-2 bomber combines LO, large payload, and long range to deliver conventional or nuclear munitions.[2] The B-2 program began during the early 1980s with the objective of penetrating sophisticated air defenses and attacking high-value and heavily defended targets. The design was modified for low-altitude operations during FSD. Additional design and test efforts have been required to integrate precision conventional munitions, including the GPS-Aided Targeting System, GAM, and JDAM.

System Description

The B-2 has a two-person crew and is powered by four General Electric F-118-GE-100 engines rated at 17,300 pounds of thrust. The significant features include the following:

- reduced electromagnetic, infrared (IR), acoustic, visual, and radar signatures
- extensive use of composite structures in the airframe
- fabrication, assembly, and finishing of parts to high tolerances to achieve stealth
- a blended flying-wing shape
- two internal weapon bays
- 44,000-pound payload
- designed to carry general purpose bombs from 500 to 4,700 pounds, mines, GAMs, JDAM, and nuclear bombs[3]
- engine inlets and exhaust shaped for radar and IR stealth

---

[2] Information on the B-2 came from CTF Highlight Summary Flight Test Production (undated, after June 1997); T&E Master Plan for the Sustainment Phase of the B-2A Advanced Technology Bomber, March 23, 1999; briefing slides from and discussions with the B-2 SPO.

[3] A JSOW capability was added post-baseline.

- features offensive avionics, including Hughes Aircraft Company's APQ-181 radar (now Raytheon's)
- automatic terrain following to 200 feet.

Programmatics

A cost-plus-incentive-fee FSD contract was awarded to Northrop in November 1981. This contract specified delivery of two ground-test articles and six flight-test vehicles. Boeing and Vought teamed with Northrop to design and build the airframe. Hughes Aircraft Company was the radar contractor, and General Electric Aircraft Engine Group developed and built the engine. After flight testing, the EMD aircraft were updated to the baseline (Block 30) configuration for use as operational assets. There was no full-rate production. Programmatic milestones included the following:

- aircraft preliminary design review: November 1982 and May 1984[4]
- CDR: December 1985
- first engine delivery: December 1986
- LRIP contract: November 1987
- first flight: July 1989
- delivery of the durability test article: September 1989
- delivery of the static test article: January 1990
- first production aircraft delivery: December 1993
- end of EMD (baseline): March 1998.

Test Program

FSD/EMD began in November 1981, with first flight in July 1989 and DT ending in March 1998. The six test aircraft made 1,013 flights, flying 5,197 hours for a total of 310 aircraft months.

Dedicated IOT&E began in October 1993 and ended in June 1997. The six aircraft made 11 flights, flying 94.7 hours for these tests.

---

[4] The second review was for the low-altitude redesign.

Combined DT and IOT&E began in July 1989 and was completed in March 1998 (see Tables A.3 and A.4). A flying test bed was used for early testing of radar and navigation systems. Test personnel were organized as a CTF, which peaked at 2,011 personnel.

The B-2 did not fully meet operational requirements during the baseline program. Deficiencies were found in the defensive management, mission planning, and terrain following systems and in LO maintainability, which affected maintenance man-hours per flying hour and the mission-capable and sortie-generation rates. These issues

Table A.3
B-2 Flight-Test Program

Event	Dates	Test Aircraft	Sorties	Flight Hours
Total FSD/EMD flight test[a]	July 1989–March 1998	6	1,013	5,197.0
Flying test bed[b]	April 1987–December 1995	1	600	3,353.0
Dedicated IOT&E	October 1993–June 1997	6	11	94.7

[a]These tests accomplished 19,897 flight-test points and 4,194 ground-test points.
[b]Avionics testing.

Table A.4
The B-2 Test Aircraft

Test Aircraft	Enter Testing	Exit Testing	Finish Build Months	Aircraft Months
AV-1	July 1989	March 1993	3	41
AV-2	October 1990	August 1995	0	58
AV-3	June 1991	March 1998	12	69
AV-4	April 1992	June 1997	0	62
AV-5	October 1992	November 1997	7	54
AV-6	March 1993	December 1995	7	26
Total				310

NOTE: The Finish Build Months column shows the months that were spent purely on finishing the building of the aircraft during the testing span. This work was necessary for testing but was known to be incomplete at entry into testing. This time is not counted in the Aircraft Months column. Also not included is the time spent installing upgraded configurations resulting from testing.

were or are being addressed in the sustainment phase, along with various upgrades. DOT&E waived requirements for full-scale live-fire testing because the program was not expected to proceed beyond low-rate production.

DT Ground Tests: Wind-tunnel and weapon-separation tests took place at Arnold Engineering Development Center (AEDC). Testing of the engine inlets was at a Northrop radio frequency range in California. A six-degrees-of-freedom model for weapon separation was used to simulate weapon drops (actual drops were made to verify the model). The fatigue article at Palmdale, California, was tested to two lifetimes. The static article was tested to 160 percent of ultimate load.

DT Flight Tests: Until 1988, the flight-test plan called for four test aircraft at 50 hours per month per vehicle. This plan was based on a planned first flight in 1987, which in fact slipped to 1989. The first six aircraft were delivered late and were incomplete, so the flight-test plan was restructured in 1988 to six test vehicles at 20 hours per month per vehicle. The first two flight-test aircraft (AV-1, AV-2) were used for air vehicle testing and did not have the full avionics suite. In-flight signature testing was done at a government facility.

Several unanticipated events during FSD/EMD affected the test program including the following:

- a major redesign and retest of aircraft components early in FSD to meet radar cross section (RCS) requirements
- a change in the primary mission from delivery of nuclear to conventional weapons
- late delivery of partially complete aircraft.

IOT&E was followed by FOT&E Phase I conducted by AFOTEC at Whiteman AFB through December 1998. FOT&E testing has focused on correction of deficiencies.

C-17 Globemaster

Mission

The C-17 provides worldwide airlift for U.S. and allied combat forces, equipment, and supplies.[5] It can deliver passengers and outsize, oversize, or bulk cargo over intercontinental distances without refueling. The aircraft can land at conventional or austere airfields or can make deliveries by airdrop. The capability for rapid, in-flight reconfiguration allows the C-17 to transition easily among its mission modes.

System Description

The C-17 is a four-engine turbofan transport aircraft. Its engines are modified versions of an earlier commercial airline engine (PW-2040). Significant features include the following

- A supercritical wing design and winglets reduce drag and increase fuel efficiency and range.
- The aircraft can be refueled in flight.
- An externally blown flap configuration, direct lift-control spoilers, and a high-impact landing gear system allow the aircraft to use small, austere airfields.
- A forward and upward thrust-reverser system provides backup capability, reduces the aircraft's ramp-space requirements, and minimizes interference with ground operations.
- The airdrop system is fully automated.
- A single loadmaster can operate the cargo door, ramp, and cargo restraint systems, and off-loading equipment does not require special handling equipment.
- The flight-control system is electronic (quad-redundant, fly-by-wire).
- The two-person cockpit has multifunction displays.

---

[5] Information on the C-17 came from T&E Master Plan for the C-17 Weapon System, dated August 1999; briefing slides and discussions with C-17 SPO; C-17 Program Office Estimate, dated June 1993; C-17 Flight Test Progress Report for Month Ending December 31, 1994.

- The mission computer integrates the majority of the avionics.
- The Onboard Inert Gas Generating System handles fuel tank inerting.
- Built-in test features reduce maintenance and troubleshooting times.

Programmatics

A fixed-price-incentive-fee full-scale development contract with two production options was awarded to McDonnell Douglas Aircraft Company in December 1985. The contract was restructured in January 1988. LRIP was approved in January 1989. The lot III production contract was awarded in July 1991, and the first flight was in September 1991.

Test Program

Several developmental problems affected the test program. Software integration was more complex than originally anticipated and increased the amount of test effort, particularly for regression testing. In an attempt to identify and solve problems on the ground, McDonnell Douglas (now Boeing) established an avionics integration laboratory and a flight hardware simulator in Long Beach, California. Tables A.5 and A.6 summarize the flight-test program and aircraft used, respectively.

FSD/EMD began in December 1985; first flight was in September 1991, and DT ended in December 1994. Six test aircraft flew a total of 1,134 sorties and 4,096 flight hours. The test program used a total of 169 aircraft months. Although the original plan had been for 69 aircraft months and 2,277 flight hours (33.0 flight hours per aircraft per month), the realized rate was 24.3 flight hours per aircraft per month.

During static testing, the wing broke at 136 percent load and had to be redesigned and retrofitted to production aircraft. The original design included a hydromechanical flight-control system, but wind-tunnel testing identified problems with it. As a result the sys-

Table A.5
C-17 Flight-Test Program

Event	Test Aircraft (no.)	Sorties (no.)	Flight Time (hrs.)	% of Flight Time
FSD flight testing September 1991–December 1994				
Planned testing				
Avionics			565	
Flying qualities			460	
Aerodynamic performance			373	
Aircraft systems			319	
Mission systems			256	
Structures			213	
Test unique			44	
Subtotal planned			2,230	54.4
Unplanned testing				
Unplanned demand			1,074	26.2
Test and work requests			763	18.6
Added tests			29	7.0
Subtotal unplanned			1,866	45.6
Total FSD flight testing	6	1,134	4,096	
DT/IOT&E June 1992–December 1994	2			
Dedicated IOT&E December 1994–June 1995	2			

NOTE: Total FSD testing included 5,623 flight-test points and 1,028 ground-test points.

Table A.6
The C-17 Aircraft Tested

Test Aircraft	Entered Testing	Exited Testing	Aircraft Months
T1	September 1991	December 1994[a]	39
P1	May 1992	December 1994[a]	31
P2	June 1992	October 1994	28
P3	September 1992	December 1994	27
P4	December 1992	December 1994[a]	24
P5	January 1993	September 1994	20
Total			169

[a]At the end of DT.

tem, was changed to a four-channel fly-by-wire system. A mechanical backup was added later, although the testing approach presumed total reliance on the electronic system. Effective use of M&S allowed live shots on a wing section to satisfy LFT&E requirements.

OT included AFOTEC's early operational assessment, in September 1988, to support the Milestone IIIA decision, and an operational assessment in January 1990. AFOTEC determined that the major risks for system development and IOT&E were software development and avionics integration. Combined DT&E/IOT&E began in June 1992. In support of a congressionally directed assessment of the C-17, AFOTEC assessed the C-17 system as adequate overall, considering its stage of development at the time, but identified range and payload, maintainability, and software maturity as risk areas.

Dedicated IOT&E began in December 1994 and ended in June 1995 and involved two aircraft. The original plan was to dedicate three aircraft months to training and eight aircraft months to IOT&E. This testing was conducted in three phases. Phase I, December 1–17, 1994, focused on C-17 cargo loading and transportability. Phase II evaluated all operations except static line paratroop drops and a "slice" of brigade airdrop demonstration. These areas were evaluated in Phase III, which was completed in June 1995.

AFOTEC's final IOT&E report included data generated during initial squadron operations and the reliability, maintainability, and availability evaluation. AFOTEC judged the C-17 to be operationally effective and suitable, meeting all key parameters and demonstrating outstanding direct-delivery capability and maintainability. Areas for improvement included formation personnel airdrop, mission computer takeoff and landing data, aeromedical evacuation capability, fault isolation, support equipment, and software maturity.

F/A-18E/F Super Hornet

Missions
The F/A-18E/F is a carrier-based multimission strike fighter derived from the F/A-18C/D. The F/A-18E/F's range, payload, and surviv-

ability have improved over those of its predecessor.[6] Missions include fighter escort, combat air patrol, interdiction, and close air support.

System Description

The F/A-18E is a single-seat and F/A-18F is a two-seat combat-capable trainer. Every part of the F/A-18E/F structure was redesigned from its predecessor. In general, the structure was enlarged and strengthened, the part count was reduced, and the use of materials and tooling were changed. Ninety-six percent of the airframe unit weight is unique to the E/F. The structure's material composition features more titanium and composites and less aluminum than its predecessor. Changes from the F/A-18C/D include the following:

- The gross landing weight increased by 10,000 pounds.
- Redesigning the forward fuselage increased its strength and decreased the part count.
- The center-aft fuselage was lengthened 34 inches to increase fuel capacity.
- Wing area increased 100 ft², and the wingspan increased by more than 4 ft, also increasing internal fuel capacity.
- The areas of the control surfaces, horizontal tail surfaces, and leading-edge extension increased.
- Unitizing reduced the part count by 42 percent from that of the C/D and reduced manufacturing costs.
- The new configuration can carry an additional 3,600 pounds of fuel internally and 3,100 pounds externally.
- The aircraft has two additional hard points for weapons.
- Incorporating such low-cost stealth features as saw-toothed doors and panels, realigned joints and edges, and angled antennas reduced the RCS.

---

[6] The test program description came from TEMP No. 0201-04 Rev. B; the dates, number of flights, and aircraft months came from the EMD flight log that the program office provided; the system description comes from various sources.

- Although the E/F and the F/A-18C/D have many common subsystems, the E/F has more powerful actuators to accommodate larger control surfaces.
- Over 90 percent of the avionics are common with the F/A-18C.
- General Electric developed the F414-GE-400 turbofan engine, rated at approximately 22,000 pounds thrust, for the new aircraft. Two engines power the E/F.

Programmatics

The Navy awarded a cost-plus-incentive-fee contract for airframe EMD to McDonnell Douglas (now Boeing) in June 1992. Northrop is the major airframe subcontractor and is responsible for the center and aft fuselage, vertical tail, and several subsystems. Three ground-test articles—static, drop, and fatigue—were built, and seven flight-test vehicles were built and flight tested in EMD.

- EMD contract award: June 1992
- CDR: June 1994
- first flight: November 1995
- fatigue testing completed: July 1998
- flight testing completed: April 1999.

Test Program

Two phases of studies and testing preceded EMD. The configuration study phase (1988 to 1991) used approximately 600 hours of wind-tunnel tests of a 12-percent scale model to define the current aero-dynamic configuration and engine performance requirements. During the pre-EMD phase (1991 and 1992), an additional 2,000 hours of wind-tunnel tests on models helped optimize the configuration and reduce design risks before EMD.

DT in EMD was conducted in several phases. During DT-IIA (November 1995 to November 1996), the static test article was used for initial wing-bending tests; the drop-test article was used for a series of landing-gear tests at increasing sink rates; the manned flight simulator was used for aircrew training and other tests; and flight testing focused on envelope expansion.

FSD/EMD began in June 1992 and ended in April 1999, with first flight in November 1995. Eight aircraft made 3,141 flights, logging 4,620 flying hours, for a total of 244 aircraft months. OPEVAL involved seven aircraft and began in May 1999 and ended in November 1999. Tables A.7 and A.8 summarize the flight-test program and aircraft used, respectively.

Table A.7
F/A-18E/F Flight-Test Program

Event	Test Aircraft (no.)	Sorties (no.)	Flight Time (hrs.)
Total FSD Flight Test November 1995–April 1999	8	3,141	4,620
Avionics			600
Flying qualities			1,890
Prop. performance			195
Aircraft systems			275
Armament			310
Structures			1,126
Carrier suitability			189
Other			35
OPEVAL May 1999–November 1999	7		

NOTE: One LRIP aircraft was used for two months in FSD testing.

Table A.8
The F/A-18E/F Test Aircraft

Test Aircraft	Enter Testing	Exit Testing	Aircraft Months
E-1	November 1995	April 1999	41
E-2	December 1995	April 1999	40
E-3	January 1997	April 1999	27
E-4	July 1996	April 1999	34
E-5	August 1996	April 1999	32
F-1	April 1996	April 1999	37
F-2	October 1996	April 1999	31
F-4	February 1999	April 1999	2
Total			244

In March 1996, early in the flight-test program, there was a wing drop incident.[7] This occurred during high-speed maneuvers and prevented the pilot from performing close-in tracking maneuvers on potential adversaries. After identifying the wing-drop problem, a Boeing-Navy team performed wind-tunnel tests and computational fluid-dynamics studies to identify the cause. The results indicated that the problem was associated with airflow separation differences between the left and right wings. Boeing and the Navy considered three solutions to the problem and implemented a change to the wing to correct the problem.

During DT-IIB (December 1996 to November 1997), static and drop testing were completed; fatigue testing began; and flight testing focused on expanding the flight envelope, initial sea trials, evaluation of the aircraft in the carrier environment, evaluation of aircraft subsystem performance, and start of EW suite and IR signature testing.

In DT-IIC (December 1997 to November 1998), engine full production qualification ground tests were completed during over 10,000 hours of testing. LFT&E was conducted using the drop-test article. Live-fire tests included analysis of previous F/A-18A/B live-fire tests and used a ground-test article for eight major live-fire tests to evaluate vulnerability of the F/A-18E/F tail, wing, and fuselage. IR signature and EW suite tests were completed. Flight testing included dynamic RCS measurements, flight-envelope expansion tests, and weapon clearance tests.

DT-IID (November 1998 to April 1999) was the TECHEVAL. Testing focused on validation and verification of production-representative weapon software functionality, EW suite testing, and testing in the carrier environment.

DT flight testing was finished at the end of April 1999, completing over 15,000 test points and clearing 29 weapon configurations for flight.

---

[7] *Wing drop* is an abrupt, uncommanded rolling motion of the aircraft during certain flight conditions.

OPEVAL (May 1999 to November 1999) involved seven production aircraft. Much of the testing was at China Lake, California, with deployments to other locations, including carrier operations.

F/A-22 Raptor

Mission

The F/A-22's primary mission is air superiority, with a secondary air-to-ground mission when equipped with JDAM.[8] Its combination of sensors, displays, weapons, and LO is designed to provide first-look, first-kill capability in all tactical environments. It will eventually replace the F-15 in the air superiority role. Its designation was changed from F-22 to F/A-22 in September 2002 to recognize its dual role.

System Description

The F/A-22 Raptor is a twin-engine, single-seat, LO, all-weather fighter and attack aircraft. The following are some of its key features:

- Advanced turbofan engines, which allow sustained supersonic cruise without afterburners, and thrust vectoring provide enhanced performance.
- Reduced radar and IR signatures and internal weapon carriage provide LO.
- The advanced integrated avionics include
  - fusion of radar, EW, and communications, navigation, and identification sensor outputs
  - long-range, active and passive, offensive and defensive sensors to improve detection and tracking

---

[8] Information on the F-22 came from the F-22 T&E Master Plan, Version 1.0, dated September 2000; Draft F-22 CARD, dated April 1999; briefing slides and discussions with F-22 SPO; From the Advanced Tactical Fighter to the F-22 Raptor, ANSER, March 24, 1998.

- multispectral, wide-aspect threat warning and tailored multi-spectral defensive response with automated and manual modes
- modular open architecture with inherent growth capability.

The operational software, which is developed and released in blocks, consists of approximately 2.1 million lines of code.

- Automation and optimized pilot interfaces provide expanded situational awareness.
- Onboard support systems with extensive integrated diagnostics and fault and failure tolerance capability provide improved supportability; accessibility has been simplified by reducing the tool set and the amount of unique support equipment.

Programmatics

The Advanced Tactical Fighter entered its DEM/VAL phase in October 1986 and completed it in August 1991. This phase produced the YF-22 (Lockheed, Boeing, and General Dynamics) and YF-23 (Northrop and McDonnell Douglas) prototypes to demonstrate airframe and engine design approaches.

In August 1991, the F-22 entered EMD, with Lockheed Martin selected as the system contractor and Pratt & Whitney as the engine contractor. The EMD contract specifies nine EMD aircraft, two ground-test articles, and a full scale pole model for RCS testing. In 1996, the Air Force Acquisition Executive, concerned about cost growth trends on the program, chartered a joint estimating team (JET) to assess the F-22's costs and schedule. The JET recommended delaying the transition to production and adding 12 months for completing avionics development. The restructured program dropped the preproduction verification aircraft. Congress imposed a cost cap on development and production that was based on the JET projections, as well as exit criteria the program had to meet before Congress would authorize the transition to production.

The EMD aircraft first flew in September 1997. In May 1998, a contract for two production-representative test vehicles (PRTVs) and a first lot of six production aircraft was awarded. A separate "program support" contract for contractor tasks not directly identifiable to a specific aircraft was let to provide cost traceability to the negotiated target price curve. In its FY 2000 appropriations bill, Congress approved redesignating the six lot I aircraft as PRTV lot II and procuring them under RDT&E funding. As of October 2002, all nine test aircraft had been delivered, and the test aircraft to be used in dedicated IOT&E were undergoing structural and electrical system modifications.

Test Program

The F/A-22 test program is representative of the most complex production aircraft testing because it combines a new, advanced-design airframe, engine, avionics, and LO features in a single, highly integrated system. This translates into additional testing complexity (see Table A.9 for a summary).

The test programs during DEM/VAL focused on allowing the two competing contractor teams to demonstrate key technologies and risk reduction. It was not a competitive "fly-off." The government did not specify the testing, and the test results were not a deciding factor in the EMD source selection. The contractors, with government participation, executed a short but intense set of flight demonstrations.

One carryover from the DEM/VAL phase is the use of a highly modified Boeing 757 flying test bed to test and troubleshoot F/A-22 avionics and software before its installation on the F/A-22. The test bed has an F/A-22 forward fuselage grafted onto its nose and a wing for mounting sensors attached to the upper fuselage, above the cockpit.

The program also used the Variable Stability In-Flight Simulator Test Aircraft, an F-16 specially configured to mimic the flying characteristics of the F/A-22. This aircraft was used to verify the flight control laws to be used in the F/A-22 flight control system.

The F/A-22 CTF consists of DT&E and OT&E test personnel from the contractors, the Air Force Flight Test Center, AFOTEC, the F/A-22 SPO, and Air Combat Command. Of the 770 personnel on the CTF in late 2001, 480, or 62 percent, were government (mili-

Table A.9
Special Requirements for F/A-22 Testing

Feature	Effects on Testing
Reduced signature design	Requires • maintaining several test aircraft in the LO configuration • obtaining and scheduling unique LO test assets • managing security considerations.
Internal weapon carriage	Requires • additional wind-tunnel characterization of flow field with bay open • an additional flight-test configuration (doors open) for performance and flying qualities.
Sustained supersonic cruise	Reduces test time with chase aircraft. Requires additional tanker support. Increases use of supersonic test airspace.
Thrust vectoring	Requires • special ground-test fixtures to control vectored exhaust gases • multiaxis force and moment instrumentation for thrust measurement • ground and in-flight performance testing • expanded flying and handling quality testing • failure modes and effects testing, particularly with respect to asymmetric actuation.
Integrated avionics	Requires • additional EMI and EMC testing • comprehensive ground and air testing of integrated system modes • collecting data on system timelines and their effect on system performance.
Sensor fusion	Requires • high-density, multispectral, integrated, and enhanced-fidelity target and threat simulation • comprehensive integrated ground-test facilities.
Highly integrated wide-field-of-regard sensors	Multiple threat and target simulators with high update rates are concurrently operated within a large field of view.
Tailored countermeasures	Requires • a target platform with representative signature • air and ground threats that appropriately stimulate the system to determine countermeasure effectiveness.

Table A.9—continued

Feature	Effects on Testing
Integrated maintenance information system and technical order data	Software intensive, paperless systems require first-of-kind DT/OT evaluations and assessments.
AFMSS and mission support element	Because these elements of the weapon system have a higher level of integration, the testing activity also must have a higher level of integration.

SOURCE: F-22 TEMP.

tary, civilian, or support contractor), and the remaining 290 were weapon system contractor personnel. Since the first EMD flight, the CTF has been collocated at Edwards AFB, California. To identify operational suitability and effectiveness issues early in testing, DT&E and OT&E testers will participate in the planning and execution of military utility testing as part of the combined DT/OT phase of testing.

Aircraft 4001–4003 are dedicated to performance, structures, flying qualities, high angle of attack, propulsion, and stores carriage and separation testing. They are equipped with flight-test nose booms and provisions for special flight-test instrumentation, such as a flutter excitation system, a center-of-gravity control system, and a stabilization recovery chute. Aircraft 4004–4009 have full avionics suites and will be used for avionics and weapon integration testing. AFOTEC will use two EMD aircraft (modified to be production representative) and two PRTVs for dedicated IOT&E. An additional production-representative spare aircraft will also be available.

For LO testing, the full-scale pole model was used to finalize the design and to assess signatures against the RCS specification. In-flight RCS testing using at least three EMD aircraft will support final signature verification. Multiple in-flight RCS tests over the test period, using at least two aircraft, will be used to verify maintenance procedures. AFOTEC test flights will be used both to verify the RCS and the indications of the RCS status from the signature assessment system.

The test program has been rebaselined several times since the beginning of EMD. To date, approximately 122 test months have been lost primarily because of late aircraft deliveries for testing. Table A.10 presents RAND's projections for the total F/A-22 flight-test program, based on the flights and flight hours through June 2001 plus the SPO's estimate of additional flights and flight hours based on the June 2001 replan.

Table A.11 presents the aircraft calendar months in flight test (after initial airworthiness testing). Delivery dates for aircraft that had yet to be delivered are as of January 2002.

Table A.12 reports the DEM/VAL and EMD schedules, along with sorties, aircraft, and rates, for the F/A-22 planned flight test program.

Table A.10
F/A-22 Flight Test; June 2001 Replan

Event	Test Aircraft (no.)	Sorties (no.)	Flight Time (hrs)
EMD flight test June 2001– January 2004	8[a]	1,658	3,680
Dedicated IOT&E April 2003–November 2003	4	428[b]	856

[a]Aircraft 4001 retired from flight test before the June 2001 replan.
[b]428 scheduled/322 effective sorties, amounting to approximately 28 aircraft months.

Table A.11
The F/A-22 Test Aircraft

Test Aircraft	Enter EMD Testing	Exit EMD Testing[a]	Aircraft Calendar Months to Complete EMD Testing
4001	May 1998	November 2000	31[b]
4002	August 1998	April 2002	44
4003	September 2000	January 2004	41
4004[c]	January 2001	August 2003	32
4005	March 2001	August 2003	30
4006	May 2001	August 2003	28

Table A.11—continued

Test Aircraft	Enter EMD Testing	Exit EMD Testing[a]	Aircraft Calendar Months to Complete EMD Testing
4007	January 2002	April 2003[d]	16
4008	March 2002	September 2002[d]	7
4009	May 2002	September 2002[d]	5
Total			234[e]

[a]SPO projections for completing all EMD testing.

[b]Note that 4001 retired before the June 2001 replan.

[c]Dedicated to climatic testing for a significant portion of this period. This time has not been subtracted.

[d]These aircraft were assigned to dedicated IOT&E training and dedicated IOT&E, either as primary or backup, and therefore considered not available for test.

[e]This total includes periods when a test aircraft was unavailable for flight test because of modifications, ground tests, software loading, etc. Some of these times were planned on June 2001, and some surfaced during execution of the program.

Table A.12
F/A-22 Planned Flight-Test Program

	First Flight	Aircraft (no.)	Sorties (no.)	Aircraft Months	Flight Time (hrs)	Rate[a]
DEM/VAL October 1986– January 1991						
(YF-22)	August 1990	2	74	5[b]	92[a]	18.4
(YF-23)	September 1990	2	50	5[b]	65	13.0
EMD[c] August 1991— January 2004	September 1997	9	1,659	234	3,680	10.7

[a]In flight hours per aircraft month.

[b]Approximate.

[c]SPO projections for completing all EMD testing.

T-45 Naval Undergraduate Jet Flight Training System

Mission

The Undergraduate Jet Flight Training System (UJFT) provides intermediate and advanced strike pilot training using an integrated ground and flight training system.[9] The flight training includes aircraft familiarization, basic instruments, airway navigation, air-to-ground and simulated air-to-air weapon delivery, aerial combat maneuvering, carrier qualifications, low-level navigation, formation flying, and tactical maneuvering. The system also supports an instructor training course. The T-45 Training System was developed to replace both the T-2C and TA-4J.

System Description

The T-45 Training System (T45TS) consists of the T-45 Goshawk aircraft; an aircraft simulator suite for both instrument and visual flight training; flight training–related academic materials, including training courses, equipment, and course materials for UJFT and instructor training; a computer-based training integration system; and contractor logistics support.[10]

The T-45 Goshawk is a tandem-seat single-engine carrier-capable jet aircraft derived from the existing BAe Hawk. The aircraft includes the Navy aircrew common ejection seat, the standard attitude heading reference system, an onboard oxygen generating system, carrier operations capability, and a training weapon delivery capability. The T-45 is fully contractor supported, including all levels of maintenance and logistics.

Programmatics

The T-45 entered EMD in 1984. The initial OT (OT-IIA) in November 1988 identified major deficiencies in aircraft handling

---

[9] Information on the T-45 and T-45 Cockpit 21 came from the *Test and Evaluation Master Plan No. 786* for the Naval Undergraduate Flight Training System (T45TS) (Revisions 5 and 6), briefing slides and discussions with the program office.

[10] Note that, for this study, we addressed only the aircraft portion of the system.

qualities, which caused several program slips while Boeing (then McDonnell Douglas) was redesigning the aircraft. OT-IIB in 1990 and OT-IIC in 1991 verified improvement in the deficient areas, and a successful OPEVAL (OT-IID) was completed in April 1994. The T45TS was determined to be operationally effective and operationally suitable.

Test Program

The T-45 was originally planned as a firm-fixed-price demonstration program that would require only relatively minor modifications. It was the first modern land-based aircraft to be modified for carrier capability. However, much of the structure of the Hawk had to be changed to make it carrier-suitable.

As a result, the program evolved into a major development effort. Because it used an existing airframe, little M&S was originally planned; this increased the flight hours necessary to test changes to the airframe. Because of the performance shortfalls and redesign, normal aeronautical and engine simulations eventually had to be developed anyway.

Relatively few government test engineers were assigned to the program. The Naval Air Warfare Center's Aircraft Division (NAWC-AD) did most of the government DT. High-angle-of-attack testing was done at the Air Force Flight Test Center. Most of the contractor's testing was done at its facility in Yuma, Arizona. One test aircraft was lost in a Class A mishap, resulting in a 12–18 month slip. Live-fire testing was not required. COMOPTEVFOR conducted the operational assessments and testing.

All OT periods (and associated flight hours) are government only, as reported in the T-45 TEMP (see Table A.13). The contractor flew hundreds of additional flight hours during DT, which the TEMP does not include. From the beginning of DT to February 1, 1994, which encompasses all DT testing and the first OPEVAL period, the total contractor and government flight testing was 1,880 flights and 1,932 flight hours. Table A.14 summarizes the information in Table A.13.

Aircraft Program Descriptions   135

Table A.13
T-45 Flight Testing

Event	Test Aircraft (no.)	Sorties (no.)	Flight Time (hrs)	Remarks
DT-IIA November 6–17, 1988	1	17	20.2	Evaluation of flying qualities, systems, and performance
OT-IIA November 17–21, 1988	1	10	13.3	Did not support LRIP; program restructured
DT-IIA follow-on February 27–March 11, 1989	1	8	9.8	Further evaluation of flying qualities and performance
DT-IIA additional follow-on June 1990	2	3	6.1	Evaluation of human factors, controls, and displays
DT-IIB November 16–December 6, 1990	2	27	27.1	Verification of corrections of prior deficiencies, suitability, specification conformance
OT-IIB December 13–20, 1990	2	20	24.1	Supported LRIP
DT-IIC July 23–October 7, 1991	2	19	20.0	Evaluation of weapon carriage, high speed flying qualities, ILS, VOR, lighting, and human factors
OT-IIC August 1–8, 1991	3	19	22.5	Testing of various training missions
DT-IID August 13–December 10, 1991 and January 3–7, 1992	2	22	7.0	Land-based catapult and arrestment
		4	8.2	Initial sea trials; terminated because of aircraft damage
	1	10	8.7	Evaluation of expanded-envelope flying qualities
DT-IIE September 1–6, 1993	2	25	35.5	Follow-on sea trials
DT-IIF September 1–November 23, 1993	6	47	52.0	High angle of attack testing at Edwards AFB
		28	30.7	Navy TECHEVAL
		12	27.3[a]	Follow-on sea trials
OT-IID October 10–November 17, 1993 and February 8–April 8, 1994	6	583	671.8	System OPEVAL, including aircraft, academics, Training Integration System, and simulators

[a]Estimated.

Table A.14
T-45 Testing Summary

Event	Test Aircraft	Sorties	Flight Time (hrs)	Remarks
DT and first OPEVAL period November 1988–February 1994	1–6	1,880	1,932	Contractor and government flight testing; includes some flights and hours shown below under OPEVAL
OT-IID OPEVAL October–November 1993 and February–April 1994	6	583	671.8	System OPEVAL, including aircraft, academics, training integration system, and simulators

T-45 Cockpit-21

Mission

The T-45 Training System provides intermediate and advanced strike pilot training using an integrated ground- and flight-training system. Undergraduate jet pilot training (UJPT) includes aircraft familiarization, basic instruments, airway navigation, air-to-ground and simulated air-to-air weapon delivery, aerial combat maneuvering, carrier qualification, low-level navigation, formation, and tactical maneuvering. Also, the system supports an instructor training course. The T-45TS was developed to replace both the T-2C and TA-4J.

System Description

The T-45 Training System (T45TS) consists of the T-45 aircraft; an aircraft simulator suite for both instrument and visual flight training; academic materials, including training courses, equipment, and course materials for UJPT and instructor training; a computer-based training integration system; and contractor logistics support.[11] The

---

[11] Note that, for this study, we addressed only the aircraft portion of the system.

first 83 aircraft delivered to the Navy were T-45As with analog cockpits. Subsequent deliveries were T-45Cs, which incorporate a digital cockpit known as Cockpit-21. To provide an early introduction to digital displays similar to those used in current fleet tactical aircraft, the Cockpit-21 avionics upgrade replaces current conventional primary flight instruments with two multifunctional displays in each cockpit.

Programmatics

Prototype testing of the Cockpit-21 upgrade began in March 1994, with the first flight the same month. Approval for fleet introduction of the T-45C was recommended in December 1998, following OT-IIIB earlier that year.

Test Program

Contractor and government flight testing ran from March 1994 to September 1998 and involved up to four aircraft (see Table A.15).

Table A.15
T-45C Flight Testing

Event	Test Aircraft (no.)	Sorties (no.)	Flight Time (hrs)	Remarks
DT-IIG March 1994– March 1995	1	165	215	Combined contractor-government testing of prototype digital cockpit installation, conducted in St.Louis and at NAWC-AD Patuxent River.
				System judged not ready for OT; schedule extended 1 year.
DT-IIIA March 6–25, 1996	1	24	33.2	Combined contractor-government testing of preproduction digital cockpit installation and verification of correction of deficiencies (VCD). Conducted at NAWC-AD Patuxent River.
				System judged ready for OT.

Table A.15—continued

Event	Test Aircraft (no.)	Sorties (no.)	Flight Time (hrs)	Remarks
OT-IIIA April 9–24, 1996	1	28	38.4	OT, including a combined DT/OT at sea test period. Judged potentially operationally effective and suitable.
DT-IIIB November 18, 1997– February 9, 1998	1	27	31.5	Conducted at NAWC-AD Patuxent River. Recommended that the T-45C (Cockpit-21) proceed to OT-IIIB.
OT-IIIB February 18– September 18, 1998	4	602	881.3	The T-45C aircraft was determined to be operationally effective and operationally suitable.

These aircraft flew 846 sorties and flew for 1,199.4 hours. Following OT-IIIB, the T-45C aircraft was determined to be operationally effective and operationally suitable.

V-22 Osprey

Mission

The V-22 weapon system is a multiservice, multimission vertical and short takeoff and landing aircraft.[12] The Marine Corps MV-22's primary mission is amphibious assault. The Air Force CV-22's primary mission is long-range infiltration, exfiltration, and resupply of Special Operations Forces. Secondary missions are land assault, medical evacuation, fleet logistics support, and special warfare. The MV-22 will replace the CH-46E and CH-53A/D in the Marine Corps inventory. The CV-22 will replace the MH-53J and MH-60G and will supplement the MC-130 in the Air Force inventory.

---

[12] We derived the information in this section from a meeting with the V-22 Program Office July 24, 2001, and from subsequent comments and input; numerous issues of Bell-Boeing's *Tiltrotor Times*; and the draft V-22 Osprey TEMP No. M960 Rev. B.

System Description

The V-22 is a tilt-rotor aircraft with rotating engine nacelles mounted on each wingtip, enabling vertical and short takeoff and landing. The nacelles rotate to the horizontal position for cruising at high speed. The tilt-rotor design combines the vertical flight capabilities of a helicopter with the speed and range of a turboprop airplane and permits aerial refueling and worldwide self-deployment. Two Rolls Royce T406-AD-400 turboshaft engines drive two 38-ft. diameter proprotors. The proprotors are connected to each other by an interconnecting shaft, which maintains proprotor synchronization and provides single-engine power to both proprotors in the event of engine failure. A triply redundant digital fly-by-wire system controls engines and flight controls. The airframe is primarily graphite epoxy composite. An integrated EW defensive suite that includes a radar warning receiver, a missile warning set, and a countermeasures dispensing system will be installed.

The Air Force CV-22 and Marine Corps MV-22 share the same basic airframe design. The CV-22 is configured for Special Operations Forces' infiltration and exfiltration operations. Unique CV-22 items include a terrain following and terrain avoidance radar system, additional fuel tanks to increase range, an additional crew position, an expanded communications suite, and the addition of a defensive systems suite to enhance survivability during penetration missions.

Programmatics

V-22 development and testing have taken place in two distinct programs. An FSD contract was awarded to a Bell-Boeing joint venture team on May 2, 1986. Six flight-test articles were planned, and five were completed. First flight of the V-22 in FSD was in March 1989. The Secretary of Defense cancelled the program on April 19, 1989, but Congress and the services continued to fund the test program incrementally until early FY 1993.

Approximately 820 hours of flight testing were completed. Two aircraft were lost: Aircraft number five crashed on its first flight in June 1991 because of a problem with its gyroscopic sensors, and aircraft number four crashed on July 20, 1992, because a section of the drive shaft failed. On October 22, 1992, the FSD airframe contract was officially terminated, and a letter contract for EMD was awarded on the same day.

The purpose of the V-22 EMD program is to design a lighter, lower cost aircraft than the FSD design and to build four production articles (aircraft numbers 7–10). An EMD engine contract was awarded in December 1992. DT and OT continued during EMD using the three remaining FSD aircraft, supplemented by EMD aircraft as they were built.

Test Program

During FSD (1986 to 1992), the V-22 T&E program concentrated on engineering and integration testing performed by the contractor. NAWC-AD conducted three periods of formal DT, with the participation of the operational test community in integrated test team activities. These tests provided early insight into the development effort.

The EMD flight-test program began by using the FSD V-22 aircraft for design support, risk reduction, and envelope expansion. A CTF, consisting of Navy, Air Force, Bell, and Boeing personnel, conducted the EMD test program, with the exception of dedicated operational test events. Testing of four EMD aircraft began in FY 1997, following the first flight of aircraft number 7 in February 1997. Integrated testing (IT) and OT used a combination of the FSD and EMD V-22 aircraft during initial assessment (IT-IIA/B/C and OT-IIA/B/C). OT-IID and subsequent tests used the EMD configuration aircraft.

The Navy conducted DT&E of the MV-22 at Patuxent Naval Air Station, Maryland. A CTF stationed at Edwards AFB, California, conducted the DT&E for the CV-22.

The Secretary of Defense certified a waiver of full system-level live-fire testing. A comprehensive series of ballistic tests (582 shots over 16 years) of critical components, major assemblies, and aircraft structures was conducted. According to DOT&E, live-fire testing was treated as "an integral part of the design process, not merely as a method of design verification." The production design was tested in 444 test firings. Live-fire testing led to a number of design changes.

A multiservice OT team under the direction of COM-OPTEVFOR conducted the MV-22 and CV-22 OT&E. Initial CV-22 OT&E culminates with OT-IIH, but six additional MV-22 and CV-22 FOT&E periods have already been identified for execution to resolve deficiencies from current OT&E, ensure that deferred OT&E events are finished, and assess P3I elements. CV-22 OT&E consists of several OT periods, designated in accordance with COMOPTEVFOR conventions as OT-IIA through OT-IIH. For the CV-22 variant, OT-IIA through OT-IIG are considered operational assessment periods, with OT-IIH functioning as the equivalent to IOT&E.

The DOT&E evaluation found the MV-22 operationally effective but not operationally suitable because of reliability, maintainability, availability, human factors, and interoperability issues. The CNO had issued a waiver from OPEVAL testing for many V-22 requirements, including combat maneuvering, cargo handling, airdrop capability, and other capabilities that affect the aircraft's operational effectiveness and suitability. DOT&E recommended further testing of these issues and of the vortex ring state phenomenon, in which the aircraft loses lift when descending at a low airspeed at too great a sink rate. One aircraft was lost because it entered a vortex ring state on an operational training mission in April 2000. When another fatal mishap occurred in December 2000, flight testing stopped for 17 months. Correction of the vortex ring state problem requires additional wind-tunnel, digital simulation, and flight testing. An additional "event-driven" flight-test program began in May 2002.

Tables A.16 and A.17 include data through OPEVAL testing in July 2000, when the V-22 was judged operationally effective but not operationally suitable. Testing continued until two fatal mishaps in 2000 caused the grounding of all these aircraft. After lengthy investigations, a new program of flight testing began in May 2002.

Table A.16
V-22 Flight Testing

Event	Test Aircraft	Sorties (no.)	Flight Time (hrs)	Remarks
FSD Flight Test March 1989– October 1992	FSD	653	763.6	
DT-IIA March 1990– April 1990	FSD	9	14.1	Evaluation of flying qualities and performance characteristics
DT-IIB November 1990– December 1990	FSD	15	16.9	Evaluation of flying qualities, performance characteristics, and shipboard suitability
DT-IIC April 1991– August 1991	2 and 4	20	29.4	Evaluation of readiness for OT-IIA Terminated on loss of aircraft 5
IT-IIA April 1993– December 1995	2 and 3	92	105.1	Evaluation of envelope expansion Exploration of structural aero-elastic and flying qualities
IT-IIB January 1996– March 1997	3	N/A	N/A	Testing supporting design, risk reduction, and pilot training
IT-IIC September 1996– May 1997	3 and 7	N/A	N/A	Pilot training and development of preliminary suitability data
IT-IID December 1996– September 1998	EMD	N/A	1,469	TECHEVAL to support OPEVAL and determine the final envelope
IT-IIE/F February 2000– December 2000	7 and 10	91	150	CV-22 testing
OT-IIA May 1994– July 1994	FSD	12	14.8	Testing in support of the Milestone II+ decision and EMD CDR, consisting primarily of ground tests and simulations

Table A.16—continued

Event	Test Aircraft	Sorties (no.)	Flight Time (hrs)	Remarks
OT-IIB June 1995– October 1995	FSD	8	10.4	Assessment of operational effectiveness and suitability; primarily ground tests and simulations
OT-IIC October 1996– May 1997	3 and 7	N/A	36.1	Assessment of operational effectiveness and suitability and support LRIP 1 decision; primarily ground tests and simulations
OT-IID September 1998– October 1998	9 and 10	63	142.6	Assessment of operational effectiveness and suitability and support LRIP 3 decision
OT-IIE, F, G, OPEVAL November 1999– July 2000	LRIP[a] 11–15	522	805	Determination of operational effectiveness and suitability of the MV-22. Assessment of all CV-22 COIs during three at-sea periods. Judged operationally effective but not operationally suitable

[a]28 aircraft months.

Table A.17
V-22 Testing Summary

	First Flight	Aircraft (no.)	Sorties (no.)	Aircraft Months	Flight Time (hrs)
FSD May 1986– October 1992	March 1989	5 (2 lost)	653	N/A	763.6
EMD October 1992– May 2000	February 1997	4 and 3 FSD	N/A	N/A	N/A
VCD June 2000 (ongoing)		8 and 10			2000 to date

Missile Program Descriptions

AIM-9X Sidewinder

Mission

The AIM-9 Sidewinder is a heat seeking air-to-air missile designed for short-range engagements.[1] It complements the medium-range AMRAAM in the fighter aircraft arsenal. The AIM-9X is a major modification of the AIM-9, which the U.S. Naval Weapons Center at China Lake, California, first developed in the 1950s. The missile has had several variants and is used on all U.S. fighter aircraft, including the F-15, F-16, and F/A-18, and will be employed on the F/A-18E/F, F/A-22, and JSF. Many allied nations also use the Sidewinder, and the Soviets and their allies copied and used its early design.

System Description

The AIM-9M, the AIM-9X's immediate predecessor, can engage targets from all aspects; its IR counter-countermeasures and background discrimination capability are better than those of its own predecessors; and it has a reduced-smoke rocket motor. Deliveries of the M model began in 1983.

The AIM-9X has improved counter-countermeasures, acquisition range, off-boresight capability, background discrimination,

---

[1] We derived the information in this section from the AIM-9X Selected Acquisition Report, December 31, 1999; an interview with the AIM-9X Joint Program Office (JPO), May 24, 2001; AIM-9X JPO (2001a) and (2001b); Sidewinder Missile AIM-9X CARD, Final Update, July 7, 2000, PMA-259; Boe and Miller (undated).

maneuverability, and day or night capability. It uses the same motor, warhead, and fuze as the AIM-9M and is of similar size and weight. Component differences include the following:

- A new airframe design and fixed forward wings reduce missile size and drag. The smaller airframe ensures that the missile will meet compressed carriage requirements for the F-22 and JSF, which have internal weapon bays.
- A control actuation system in the rear of the missile provides thrust vectoring and moveable fins for missile maneuvering. The tail control is a new development item.
- An improved seeker dome provides improved IR properties.
- A servo-controlled sensor assembly provides high off-boresight slaving.
- A 128 by 128 focal plane array (instead of the AIM-9M's single-element IR detector) produces a digital image for use as a tracker algorithm and enables the system's robust IR counter-countermeasures. This component was developed for Advanced Short-Range Air-to-Air Missile (ASRAAM).
- An electronics unit provides the AIM-9X with digital signal processing and tracking capability, IR counter-countermeasure logic, fly out guidance, and field reprogrammability. The guidance system is a new development item.

A related program, the Joint Helmet Mounted Cueing System (JHMCS), is being developed concurrently but separately from the AIM-9X. The JHMCS requires a modified helmet and new aircraft hardware and software. The JHMCS moves the head-up display to the helmet and enables slaving aircraft weapons and sensors, such as the AIM-9X, to head movements. Neither program is required for the other program to proceed, but AIM-9X and JHMCS are integrated and tested together as much as possible.

Programmatics
The AIM-9X is a joint Navy–Air Force program, with the Navy as lead service. AIM-9X is a CAIV program, trading off performance and cost to achieve a best-value solution. A two-year concept-

exploration phase preceded formal development. In this phase, the services developed five key performance parameters for the missile:

1. the ability to operate during day or night
2. the ability to operate over land or sea in the presence of IR countermeasures
3. weight, size, and electrical compatibility with all current U.S. fighters and the F-22
4. the ability to acquire, track, and fire on targets over a wider area than the AIM-9M can
5. a high probability that a missile launched will reach and kill its target.

The acquisition strategy involved a competitive two-contractor DEM/VAL phase, then downselection to one EMD contractor. Hughes and Raytheon were awarded DEM/VAL contracts in December 1994 to develop preliminary system designs and to conduct seeker demonstrations. DEM/VAL was completed in June 1996. After the 18-month competitive DEM/VAL program and evaluation of EMD and LRIP proposals, Hughes (now part of Raytheon) was selected to be the prime contractor for the AIM-9X missile development in December 1996. The contractor has total responsibility for system performance, including development, production, and lifetime maintenance support. The acquisition philosophy is intended to motivate the contractor to achieve cost and quality objectives by including both a 10-year missile warranty and award fees for reducing missile production costs. The EMD effort was scheduled to end in 2002. A 17-year production period is planned to buy a total of approximately 10,000 missiles.

The use of several existing components from the AIM-9M, including the warhead, rocket motor, and fuze, during AIM-9X development helped reduce technical risks. The AIM-9X design also includes such nondevelopmental items as the airframe and the engine control system, which the Air Force had previously developed and tested. These risk-reduction hardware programs demonstrated that a tail-controlled missile with small fins would have better performance than the AIM-9M.

Other technical risks that might have affected EMD were reduced during the 18 months of DEM/VAL; these included the seeker and tracker software, new components (such as the sensor and the guidance and control system), and other critical systems. Because of the success of the DEM/VAL program, both the contractor and the program manager considered the AIM-9X missile's overall technical risk to be low when it entered EMD. Nevertheless, there were difficult areas of development, with some technical risk, such as the development of guidance and control software for use in a countermeasures environment.

Test Program

Table B.1 summarizes the overall test program (as of June 2002), which is described in greater detail in the paragraphs below.

Table B.1
AIM-9X Testing

Event	Guided Launches	Remarks
DEM/VAL December 1994– June 1996		Two competitors

DT-IIA
January 1997–
August 1998
	0	More than 50 captive-carry
		  sorties
		Ground launch of
		  preprogrammed test vehicle

DT-IIB/C September 1998– August 2001	9	More than 300 captive-carry sorties 16 SCTV launches
DT-IID (TECHEVAL) March–December 2001	3	More than 75 captive-carry sorties
OT-IIA September 1999– July 2000	5	
DT Assist August– December 2001	4	Concurrent with TECHEVAL
OT-IIB (OPEVAL) August 2002– May 2003	22	

DT included many different types of test missions and missile configurations to clear the launch platform flight envelope; to demonstrate the missile system's performance and flight worthiness; and to test the integration of missile, launcher, aircraft, and JHMCS. The missile configurations are described below:

- **Instrumented Round with Telemetry (IRT).** The AIM-9X IRT was designed to evaluate the captive-carriage environment and its effect on the AIM-9X. Approximately 10 missions were required for each aircraft type. Missile strain gauges, accelerometers, and thermocouples gathered environmental data during the IRT test missions. The detailed test aircraft matrix included a broad range of test conditions, environments, and configurations. The IRT missions took place at the beginning of the test program to allow early modification and improvement of system reliability before the much-longer captive-carry seeker development phase of the test program. The IRT flight-test phase for the F-15C and F-18C/D was completed by August 1999.

- **Separation Control Test Vehicle (SCTV).** The first launch of the AIM-9X SCTV was in March 1999. This vehicle was designed to verify safe separation of the missile in different launch conditions from AIM-9X–capable aircraft stations on the F-15C and F-18C/D. For safety and to verify predicted missile performance, dynamic pressure and g forces were built up gradually.

  Before a guided launch could take place, a successful SCTV launch of equal or greater safe-separation difficulty was required. The F-15Cs and F-18C/Ds made 16 launches to verify the missile's envelope.

  In addition to flight envelope clearance, SCTV shots provided guidance and control data to validate models and simulations. Photo chase aircraft, test aircraft modified to carry high-speed separation cameras along with ground cameras, captured the majority of the safe-separation data. Missile fly-out was verified primarily through missile telemetry and radar and optical tracking.

- **Engineering Development Missile (EDM).** The AIM-9X EDMs were built as either a free-flight version (guided launch) or a captive unit, the EDM captive-test unit (EDM-CTU). Numerous captive missions preceded each guided launch. The EDM-CTU was designed to verify missile seeker tracker performance, verify software performance, and validate models and simulations. The EDM-CTU was subjected to multiple environments, background conditions, lighting conditions, countermeasures, acquisition ranges, boresight and aspect angles, and targets. The missions included autonomous missile acquisition and tracking, radar slaving, and slaving with the JHMCS in maneuvering and nonmaneuvering environments. The seeker test matrix was divided between the test centers to maximize the data gathered. Data collection included aircraft display videos (head-up display, radar, etc.) and the missile seeker image. Typically, the seeker image was recorded onboard the aircraft and telemetered to the control room. In addition to providing seeker performance data, EDM-CTUs demonstrated and verified weapon system integration between the AIM-9X, aircraft, launch rail, and JHMCS. The EDM-CTU also gathered captive-carriage reliability data. The majority of EDM testing was completed by June 2000.
- **Production Representative Missile (PRM).** The AIM-9X PRM, like the EDM, consists of both a free-flight version (guided launch) and a captive unit, the PRM-CTU. The PRM has improved hardware for greater image-processing capability compared to the EDM. Testing and data collection are almost identical to the EDM process. This phase began in March 2000.

In EMD, DT&E consisted of three dedicated phases, DT-IIA, DT-IIB/C, and DT-IID. To the extent practical, the testing involved production-configured missiles built on production tooling. Extensive M&S, captive-carry flights, and live firings verified achievement of required performance. The three phases were

- **DT-IIA** (January 1997–August 1998) focused on risk reduction through collection of IR imagery for seeker development and development and refinement of M&S. Wind-tunnel testing supported the six-degrees-of-freedom AIM-9X model. Over 50 captive-carry flights and a ground launch of a preprogrammed control test vehicle were conducted.

- **DT-IIB/C** (September 1998–August 2001) included laboratory testing, over 300 captive-carry flights, 16 SCTV launches, and nine guided launches. The guided launches focused on demonstrating missile operation and performance, the aircraft and launcher interface, and validation of M&S results.
- **DT-IID** (March 2001–December 2001) served as the AIM-9X TECHEVAL and included laboratory testing, three guided PRM launches, and approximately 75 captive-carry sorties.

The OT of the AIM-9X before Milestone III likewise consisted of three phases:

- **OT-IIA** (September 1999–July 2000) was an operational assessment of potential operational effectiveness and suitability before the LRIP decision. Five EDM launches were conducted to accomplish mutual DT and OT objectives. In addition, six PRMs were flown in a captive carry reliability assessment from August 2000–September 2002.
- **DT Assist** (August 2001–December 2001) was a second phase of operational assessment, concurrent with the final months of TECHEVAL, involving captive carry evaluations and four PRM launches.
- **OT-IIB** (August 2002–May 2003) will be the AIM-9X OPEVAL. Navy and Air Force personnel conducted these operations, which were scheduled to fire 22 PRMs. COM-OPTEVFOR and AFOTEC used M&S to evaluate live-fire performance and performance under conditions when live operations are not practical.

The AIM-9X paid for initial integration and certification of the AIM-9X on the F-15C for the Air Force and the F-18C/D for the Navy as part of the development program. Additional platform programs—the F-22, F-16, F-18E/F, and other aircraft—will pay for their own certification.

AIM-120 Advanced Medium-Range Air-to-Air Missile

Mission

AMRAAM is an all-weather, radar-guided, air-to-air missile that replaced the AIM-7 Sparrow.[2] The F-15, F-16, and F/A-18 currently carry this missile worldwide, and the F/A-18E/F, F-22, and JSF will carry it for use against air threats in a variety of weather and electronic combat conditions. The U.S. Marine Corps Complementary Low-Altitude Weapon System will use AMRAAM in a surface launch role. FMS platforms include the German F-4F, the Swedish Gripen and Viggen, the United Kingdom Tornado and Sea Harrier, and the Norwegian Advanced Surface-to-Air Missile System (a ground-launched version of AMRAAM). AMRAAM is capable of intercepting maneuvering and all-aspect targets within and beyond visual range and allows a single-launch aircraft to engage multiple targets with multiple missiles simultaneously.

System Description

The AMRAAM weapon system includes the missile, launcher, the fire-control system, supporting aircraft avionics, and aircrew interfaces.

AMRAAM can be either rail or ejector launched and interfaces electrically with the platforms via the launch stations. The rail launcher permits firing from wing-mounted stations and is compatible with both AMRAAM and AIM-9. The ejector launcher, which permits firing from F-15 and F/A-18 fuselage stations, is compatible with both AMRAAM and AIM-7.

---

[2] We derived the information in this section from Mayer (1993), information from the AMRAAM JPO (AAC/YAF), October 17, 2001 through November 18, 2002; Advanced Medium Range Air-to-Air Missile TEMP, March 2002.

The aircraft normally updates the missile in-flight through a data link; however, the missile design does not absolutely require this interface for successful operation at shorter ranges or against non-maneuvering targets.

AMRAAM system has four guidance modes:

- command-update data link at longer ranges, with active terminal guidance
- inertial guidance with active terminal if command update is not available
- active terminal with no reliance on the aircraft's fire-control system at distances within the seeker's acquisition range
- active radar with home-on-jam during any phase of flight.

The missile's active radar permits the launch aircraft to engage multiple targets and to use "launch and leave" tactics. The AMRAAM is propelled by a solid-fuel, reduced-smoke rocket motor.

Programmatics

The AMRAAM acquisition strategy involved a two-contractor DEM/VAL phase with downselection to a single FSD contractor. A leader-follower approach was implemented during FSD to promote competition by the fourth production lot. The AMRAAM began FSD in December 1981 with a fixed-price-incentive contract award to Hughes Aircraft Company. In July 1982, Raytheon won the second source contract. Milestone II approval was granted in September 1982. At OSD direction, the AMRAAM Producibility Enhancement Program (APREP) was initiated to lower the production cost of the missile by identifying and incrementally redesigning high-cost components. Congress imposed caps on both FSD and procurement cost. Milestone IIIA was approved in June 1987 and full-rate production (Milestone IIIB) in April 1991.

The AIM-120B missile is the result of the APREP. Improvements include a new digital processor, field reprogrammable memory, and insertion of large-scale integrated circuit and very-large-scale integrated circuit electronic components. The AIM-120B was introduced late in lot 6.

A P3I program developed the AIM-120C to provide continuous improvement in missile performance, this program proceeded in three phases, each with its own hardware and software development plan:

- P3I Phase 1 developed the AIM-120C-3, using the APREP program as a baseline and including compressed-carriage and electronic counter-countermeasure enhancements. Testing began in October 1993. Phase 1 flight testing was complete in December 1994. Production was shifted to the AIM-120C-3 in Lots 9–10 (1997–1999).
- P3I Phase 2 began in June 1994 and improved on the electronic protection and enhanced weapon effectiveness of the phase 1 (AIM-120C-3) missile. Two software tape cut-ins in lots 9 and 11 improved electronic protection. Lot 11 (AIM-120C-4) included an improved warhead, and the rocket motor for lot 12 (AIM-120C-5) was more than 5inches longer than that of the phase 1 missile. Lot 13 included a quadrant-sensing target detection device, yielding the AIM-120C-6.
- The P3I Phase 3 EMD program began in October 1998 with the objective of further improving guidance and electronic protection and developing the AIM-120C-7 variant.

In summary, the following are the AMRAAM production lots and variants produced in them:

- Lots 1 through 5 and early lot 6 produced the AIM-120A. This basic AMRAAM variant cannot be reprogrammed.
- Lots 6 through 8 produced the AIM-120B, which is reprogrammable, for the Air Force and Navy. This variant also continued to be produced for FMS through lot 14 (as of FY 2000).
- Lots 9 and 10 produced AIM-120C-3, which has clipped wings and fins to allow it to fit inside the F-22 weapon bay and incorporates electronic counter-countermeasure enhancements.
- Lot 11 produced the AIM-120C-4, which has an improved warhead.

- Lot 12 produced the AIM-120C-5, which features an extended rocket motor. Note that this variant continued to be produced for FMS customers through lot 15.
- Lots 13 through 15 produced the AIM-120C-6 (FY 1999–2001 funding), which has a new target detection device.
- Lot 16 and subsequent lots will produce the AIM-120C-7, which will have the same warhead, target detecting device, rocket motor, and control section as the AIM-120C-6 but will involve software changes and hardware modifications to the guidance section.

Test Program

Table B.2 summarizes the overall test program, which we discuss in more detail in the following paragraphs.

Table B.2
AMRAAM Testing

Event	Live Firings	Remarks
DEM/VAL November 1979– December 1981	7	One STV, three CTVs, and 3 GTVs were fired.
FSD (DT&E/IOT&E) October 1986–June 1990	100	Ninety AAVIs, 4 AAVs with warheads, and six SCTVs were fired.
AF FOT&E (1) February 1990–May 1993	37	The AFOTEC report supported full-rate production.
Initial Navy OT-IIIA January–July 1991	6	COMOPTEVFOR concluded that the missile was potentially operationally effective and suitable.
Navy OT-IIIB (OPEVAL) September 1991–May 1994	29	COMOPTEVFOR concluded that the missile was partially operationally effective and suitable and supported IOC on the F/A-18.
AF FOT&E (2) June 1993–March 1996	39	Used lot 4, 5, 6, and 7 missiles to evaluate AIM-120A software in electronic attack, chaff, and multiple-target scenarios. Testing also included an initial evaluation of the AIM-120B production software and a 7,800 hour CCRP. At least one ACE (~39) was flown for each launch.
P3I Phase 1 DT&E October 1993–October 1995	12	Two SCTVs, one JTV, and nine AAVIs were fired, and 36 development and preflight ACE missions were flown.

Table B.2—continued

Event	Live Firings	Remarks
AF FOT&E (3A) August 1996–August 1999	26	Verified the operational effectiveness and suitability of the AIM-120B/C-3 hardware and software; verified the correction of deficiencies; and completed deferred OT&E. Testing include 26 AAVI flights, 3,712 hours of captive-carry testing for CCRP, and 25 ACE missions.
P3I Phase 2 DT&E June 1994–November 2001	6	Included Phase 3 risk-reduction testing. Testing included 54 ACE flights; digital and hardware-in-the-loop simulations; and launches of one SCTV, 13 AAVIs, and one AAV (with warhead).
FDE (3B) September 2000–[a]	21	Testing included 16 AAVIs from lots 12 and 13, five captive-carry reliability vehicles (reconfigured lot 8 missiles), and a CCRP. Ten ACE missions supported the launches.
P3I Phase 3 DT&E February 2002–August 2003	13	Testing included eight AAVIs and one reliability test vehicle, seven ground and five flight ITV missions, one IMV, eight prelaunch ACE missions, and 30 ACE missions to support software development.

[a]Ongoing at reporting time, with an estimated completion date of August 2002.

Demonstration and Validation. During DEM/VAL, Hughes test-fired 38 instrumented measurement vehicle (IMV) and captive-load vehicle (CLV) missions from F-14, F-15, and F-16 aircraft and approximately 46 seeker test unit (STU) missions; one separation test vehicle (STV); one preshot guided test vehicle (GTV); and, finally, three GTV productive missions. Raytheon fired 31 IMV/CLV missions from F-14, F-15, and F-16 aircraft; 15 RAYSCAT missions,[3] two preprogrammed CTV launches, six preshot GTVs, 31 IMV/CLV missions, and one GTV. At the direction of the SPO, Hughes fired three more GTVs after DEM/VAL was over, using leftover DEM/VAL assets.

Full-Scale Development. Combined DT&E/IOT&E took place throughout the FSD phase, using both FSD and lot 1 produc-

---

[3] The RAYSCAT was a Raytheon test vehicle that was used to collect seeker waveform, cluster discrimination, and clutter data.

tion missiles. The combined DT&E/IOT&E began in October 1986 and was complete in June 1990 (the final FSD launch was in January 1989) and consisted of 79 DT&E and 21 IOT&E flights. This approach was used to achieve an earlier initial operational capability (IOC). Testing included 90 AMRAAM Air Vehicle Instrumented (AAVIs) missiles, four AMRAAM Air Vehicles (AAVs) (for warhead shots), and six Separation Control Test Vehicles (SCTVs).

AMRAAM Producibility Enhancement Program. The APREP Block I effort was designed to reduce unit production costs while maintaining baseline performance. The program introduced alternative subsystem and component designs, vendors, and production techniques, with production cut-ins occurring in lots 3, 4, and 5. APREP Block I flight testing occurred between September 1991 and March 1992. Program objectives were to demonstrate form, fit, function, and interface compatibility with the next-higher level of assembly to ensure that introduction of these projects would not degrade system performance. Tests of AMRAAM captive equipment (ACE) and simulations were conducted to demonstrate seeker performance.

APREP Block II. APREP Block II flight testing began in February 1992 and ended in December 1993, with a total of 62 ACE flights for hardware validation and software development. Nine Block II–configured missiles (AAVIs) from lot 3 were launched.

FOT&E Phase 1. AFOTEC conducted FOT&E(1) from February 1990 to May 1993, including evaluating the following:

- fixes incorporated into AIM-120A lot 2 missiles, through a captive-carry reliability program (CCRP)
- six Desert Storm missiles that had operational captive-carry time
- the effectiveness of lot 2 and 3 missiles in the operational environment
- lot 4 software capabilities, including CCRP
- the effectiveness of lot 2 through 4 missiles in multiple target, electronic attack, chaff, and maneuvering target scenarios.

AFOTEC launched 37 missiles during FOT&E Phase 1.

Initial Navy OT&E (OT-IIIA). The Navy conducted OT-IIIA from January 1991 to July 1991 to support a Milestone IIIB full-rate production decision. F/A-18C/D aircraft fired six lot 2 AIM-120A missiles.

OT-IIIB Navy OPEVAL. To support the missile's introduction into its F/A-18C/D fleet, the Navy conducted an OPEVAL of the AIM-120A from carriers off Point Mugu, California, as well as other test sites, under varying environmental conditions. The aircraft captive carried and fired 29 missiles, 17 of which were equipped with warheads. The Navy also conducted an extensive CCRP, including carried arrested landings and catapult launches.

FOT&E Phase 2. The U.S. Air Warfare Center (now the 53rd Wing) conducted FOT&E(2) from June 1993 to March 1996 to further test AMRAAM's operational capabilities. These tests evaluated the missiles operational effectiveness and suitability in tactically realistic scenarios. Tests included an extensive CCRP and an initial evaluation of the first production software for the AIM-120B. Lot 4, 5, and 6 missiles were used to evaluate improved AIM-120A software in electronic attack, chaff, and multiple-target scenarios and to identify operational capabilities and deficiencies.

P3I Phase 1. Testing of P3I Phase 1 (AIM-120C-3) began in October 1993 and included wind-tunnel testing at AEDC and IMV flights to quantify the F-15 and F-18 environments, with an F-15 and F-16 each launching one SCTV. ACE (a total of 36 productive missions) and simulations accommodated software testing. Nine AAVIs were launched to confirm that the improvements did not compromise baseline performance and three to test the improvements themselves.

P3I Phase 2. Phase 2 testing included 54 ACE flights, one SCTV to test the new control section, and one AAV to test the AIM-120C-4 configuration with the new warhead for lot 11. Thirteen AAVIs were launched to test the upgraded control activation system and the 5-inch rocket motor from lot 12. AAVI launches also tested software tape 7A, which was cut in to lot 9; 7B, which was cut in to lot 11; and 13C, which was cut in to lot 13. Four of the 13 AAVIs tested the new quadrant-sensing target-detection device for lot 13.

FOT&E Phase 3A. The Air Force's 53d Wing and the Navy's Air Test and Evaluation Squadron 9 (VX-9) conducted FOT&E(3A)

from April 1996 to June 2000 using operationally realistic scenarios at the Eglin Gulf Test Range, Eglin AFB, Florida; the White Sands Missile Range, New Mexico; the Utah Test and Training Range; and the NAWC-WD Sea Range, Point Mugu, California.

The testing served to verify the operational effectiveness and suitability of AIM-120B/C hardware and software updates,[4] to verify that deficiencies had been corrected, and to complete deferred or incomplete OT&E. The tactical scenario for each launch, coupled with threat-representative targets and electronic-attack conditions, ensured operational realism. The team launched a total of 26 missiles: six for tape 7, revision 6, profiles; three for tape 5, revision 3, profiles; and 10 for tape 7, revision 7. The Navy launched the remaining seven missiles for its own profiles. The CCRP used ten of the missiles. ACE missions and preflight simulations preceded each launch to examine additional missile capabilities (25 productive missions).

Force Development Evaluation (FDE) 3B. FDE(3B) began in September 2000. The test plan incorporates the following:

- live launches for lot verification
- periodic live launches for AIM-120 software validation and regression
- captive carry testing, suitability analysis, and live launches of projected hardware modifications
- ACE missions and computer simulations for further validation and evaluation of missile capabilities and performance, in response to inquiries from field users.

The test program consists of 21 launches of missiles from lots 12 and 13; reconfiguration and launching of five (included in the 21 launches) captive-carry reliability vehicles from lot 8, which were reconfigured and fired for additional validation. COMOPTEVFOR will conduct a CCRP to establish a baseline for the F/A-18E/F. Ten ACE runs will support the live launches.

---

[4] The items of interest were the hardware updates for lots 7–10 hardware; software update tape 7, revision 6, and tape 7, revision 7, for the AIM-120C; and software update tape 5, revision 3, for the AIM-120B.

P3I Phase 3. Phase 3 system testing began in February 2002 with ACE flights to support software development and is planned to be complete in about August 2003. DT&E objectives will support OT testing. Verification of tactical performance for the AIM-120C-7 test program will rely heavily on captive flight testing (ACE) and simulation to augment the limited number of missile firings. Raytheon will use its SIMFAX hardware-in-the-loop facility to conduct multiple test runs in a simulated flight environment using actual missile hardware and software for software development, for validation of hardware models used in the Tactical AMRAAM Simulation, and for preflight risk assessments before ACE flights and missile firings.

A planned 38 ACE captive-carry missions will provide data to support model and simulation validation and to obtain tactical performance verification data. Of these missions, 30 are for software development and to test the rehosted high-order-language software.

Eight AAVI firings (and one backup) are planned for collecting data to support model and simulation validation and to obtain end-to-end missile performance verification in a realistic environment. Eight preflight ACE missions will occur before each AAVI mission. Also, an IMV will be used early in the test program to measure vibration levels to expose deficiencies in chassis design before moving to the final design. A reliability test vehicle will be used to measure shipboard shock for deployment of the AIM-120C-7 in fleet operations.

Joint Air-to-Surface Standoff Missile

Mission
JASSM is a precision standoff weapon for attacking heavily defended, high-priority targets.[5] JASSM can be launched beyond the range of enemy air defenses and can strike fixed or relocatable targets.

---

[5] We derived the information in this section from the Lockheed Martin JASSM Web site http://www.jassm.com/; an interview with the JASSM Program Office on May 30, 2001, and subsequent communications.

System Description

JASSM is 168 inches long. Its major manufactured parts include a composite shell, fore and aft covers, tail, wings, fuel tanks, engine, and warhead. The vehicle has an LO design, and each missile is tested for its radio frequency signature. The tail and wings fold next to the body before deployment to reduce space. The 1,000-pound class warhead can penetrate hard targets, works with an impact or delayed fuze, and is compliant with Navy requirements for insensitive munitions. The missile is guided by a jam-resistant GPS and ring laser gyro inertial navigation system (INS) with an IR seeker and pattern matching autonomous target recognition system that provides aimpoint detection, tracking, and strike in the terminal phase. The control unit uses commercial electronics.

Programmatics

JASSM is the successor to the terminated Triservice Air-to-Surface Standoff Missile (TASSM) program. Two 24-month program definition/risk-reduction (PD/RR) cost-plus-fixed-fee contracts totaling $237.4 million were awarded to McDonnell Douglas and Lockheed on June 17, 1996. The PD/RR phase was completed in October 1998.

After the open competition in the PD/RR phase, Lockheed was selected as EMD contractor in April 1998, and a planned 54-month EMD program began in November 1998. The Defense Acquisition Board granted approval for LRIP in December 2001.

Test Program

Table B.3 summarizes the tests conducted for this program. The paragraphs below describe these in greater detail.

JASSM is an acquisition reform program with no government-directed DT. The contractor is responsible for planning and executing DT. The program progresses directly from contractor DT&E to OT&E. DOT&E and AFOTEC were involved during the request for proposal stage of the program to lay the foundation for data collection during contractor DT&E. Lockheed's test program (using

Table B.3
JASSM Testing

Event	Guided Launches	Launch Platform	Remarks
DT-1 January 2001	0	F-16	Inert round, seeker guides to target
DT-2 April 2001	1	F-16	Destroyed air defense target
DT-3 May 2001	1	B-52	Destroyed concrete bunker
DT-4 July 2001	1	B-52	Fuse did not arm, did not detonate
DT-5 September 2001	1	F-16	Arming problem, did not detonate
DT-6 November 2001	1	F-16	Destroyed hardened bunker
DT-7 December 2001	1	B-52	Destroyed hardened bunker
DT-8 April 2002	1	F-16	Penetrated hardened bunker
DT-9 July 2002	0	F-16	State of the art anti-jam GPS feature
DT-10 September 2002	0	F-16	Successful navigation in intense jamming environment
OT-1[a] April 2002	1	F-16	
OT-2[a] June 2002	0	F-16	Two inert rounds
OT-3[a] August 2002	1	B-52	
OT-4[a] September 2002	2	B-52	
OT-5[a] January–March 2003	2	B-52	
OT-6[a] March–April 2003	1	B-52	One live, one inert launch

[a]As of reporting time, these were planned events.

production-representative systems early in the DT program) supported data collection for combined developmental and operational test demands and will lead to an independent government IOT&E upon certification of readiness for OT by the JASSM Program Manager.

Ground tests included wind-tunnel testing of missile carriage and separation characteristics; signature tests; hardware-in-the-loop tests to simulate flight; simulation of the missile's seeker performance during autonomous navigation and terminal flight; and all-up round sled test of missile lethality against hard targets.

Flight tests included using jettison and separation test vehicles to verify safe separation from F-16 and B-52 aircraft; captive-flight functional testing and captive-carry reliability testing to verify flight readiness of missiles during prelaunch, simulated launch, and autonomous en route navigation; assessment of carrier suitability for JASSMs on F/A-18E/F aircraft during catapult launches and arrested landings; and flight test of all-up rounds to verify end-to-end system performance.

The basic premises of the DT program are that test articles will be nearly production representative and that most of the DT test events should be end-to-end system evaluations (mission planning through target damage) in operationally realistic employment scenarios using live warheads.

The main purpose of the DT program is to verify the JASSM system performance specifications. The system performance specification values are driven by three key performance parameters: expected minimum number of missiles needed to kill a target, missile operational range, and carrier operability.

Early in the EMD program, Lockheed conducted a risk-reduction free-flight test using a prototype PD/RR air vehicle. Before the end-to-end system test, two controlled flight tests were conducted at Eglin AFB, Florida, to determine the missile's aerodynamic characteristics. Lockheed successfully performed its first DT at White Sands Missile Range, New Mexico, January 19, 2001. It was the first flight using a seeker to guide to a target. JASSM flew eight powered DT flight tests through April 2002, including destruction of hardened targets. JASSM is planned to have ten DT and ten OT missile shots. The reduced number of shots (compared to other, similar programs) is due in part to the combined DT and OT test teams.

AFOTEC will be the lead OT agency for the JASSM program, with OPTEVFOR being the participating OT agency. OPTEVFOR will conduct such Navy-unique testing as FA-18E/F launch and shipboard operability during its FOT&E program.

Joint Direct Attack Munition

Mission

JDAM is an accurate, all-weather low-cost guidance kit for current inventory 2,000- (Mark 84/BLU-109) and 1,000-pound (Mark-83/BLU-110) bombs.[6] JDAM provides highly accurate weapon delivery in any "flyable" weather. After release, JDAM can receive updates from GPS satellites to help guide the bomb to the target. JDAM is a bomb-on-coordinates system that navigates from release to the planned target coordinates. JDAM provides the user with a variety of targeting options, such as preplanned targeting using the Air Force Mission Support System (AFMSS) or the Navy's Tactical Automated Mission Planning System (TAMPS), sensor targeting, or in-flight retargeting using onboard sensors and manual data entry. JDAM also allows multiple target engagements on a single-pass delivery.

System Description

The guidance kit includes an INS augmented by GPS updates and a tail control system. Target coordinates and other guidance data are passed to the weapon through a MIL-STD-1760 interface from the delivery aircraft during weapon initialization. Upon release, autonomous guidance is initiated using INS data only. After the JDAM GPS receiver acquires the GPS satellites, precision GPS position and velocity data are used to refine the INS data. The guidance unit attaches to the bomb and, through controlled tail fin movements, directs the bomb to the target. The JDAM is to be integrated on the B-1B, B-2, B-52H, F-14B/D, F-15E, F-16C/D, F-18C/D, F-18E/F,

---

[6] We derived the information in this section from the February 2000 TEMP and AFOTEC reports.

F-22, F-117A, and AV-8B aircraft. The B-52 and F/A-18C/D are the threshold aircraft.

Programmatics

JDAM is a joint Air Force–Navy program; the Air Force is the lead service. JDAM is an ACAT identification program under OSD oversight. Because JDAM was selected as a defense acquisition pilot program, many normal procurement requirements were reduced in favor of using "best commercial practices." Government management, oversight, and decision processes were also streamlined.

Four contractors were involved in pre-EMD activities. The development program had two phases. The Phase 1 EMD effort began in April 1994 and involved two competing contractors, McDonnell Douglas and Martin Marietta. The primary focus of Phase 1 EMD was to reduce manufacturing risks and the projected average unit production price by having each competitor develop its design through CDR. The Phase 2 EMD effort began in October 1995 with the selection of one contractor, McDonnell Douglas. The Phase 2 EMD effort completed system development with emphasis on development and OT. The Phase 2 EMD contract was a cost-plus-award-fee contract valued at approximately $102 million with a period of performance from October 1995 to February 1999. The Phase 2 EMD contract also included minimum and economic order quantity options for production lots 1 and 2.

Test Program

Table B.4 summarizes the test program, which we discuss in greater detail in the following paragraphs.

A test IPT that included representatives from the joint program office, 46th Test Wing, NAVAIR, and Boeing, the system contractor, managed DT. Representatives from the OT activities were also involved early in DT.

Testing in EMD Phase 1 consisted of wind-tunnel testing, digital simulation, fit checks, use of weapon simulators to test interfaces between the aircraft and the weapon, supportability analyses, and

Table B.4
JDAM Testing

Airframe	Event	Units Fired	Remarks
F-18C/D Mark 84	Safe separation 1–4Q FY1996	27 STVs	Wind-tunnel and safe-separation testing
	DT&E 1Q FY1997–3Q FY1998	48 GTVs	Captive carry and release, carrier suitability, aircraft integration, and mission planning
	DT/OT-IIA July 1998–October 1998	14 GTVs	Captive carry and release
			Flew 14 hours total with a restricted flight envelope because of problems with the tail actuator subsystem
			Concurrent DT evaluated tail actuator subsystem fixes
	OT-IIB (OPEVAL) November 1998–March 1999	55 GTVs	Independent phase of OT; tested all weapon and fuze combinations
			Flew approximately 300 hours (188 sorties)
			Verified correction of tail actuator subsystem problems
	DT/VCD 1–3Q FY2000	10 GTVs	
B-52H Mark 84	Safe separation 3Q FY1996	20 STVs	Wind-tunnel and safe-separation testing
	DT&E 2Q FY1997–2Q FY1998	16 GTVs	
	DT/OT-IIA 4Q FY1998	16 GTVs	
	OT-IIB IOT&E 1Q FY1999–4Q FY1999	40 GTVs	Independent phase of OT
			Tested all weapon and fuze combinations
			19 sorties
F-16	Safe separation 3Q FY1996 F-16 MK84	25 STVs	Wind-tunnel and safe-separation testing
	DT&E 2Q FY1997–2Q FY1998	64 GTVs	

Table B.4—continued

Airframe	Event	Units Fired	Remarks
	DT/OT-IIA 4Q FY1998	2 GTVs	
	OT-IIB IOT&E 1Q FY1999– 4Q FY1999	2 GTVs	Independent phase of OT Tested all weapon and fuze combinations 19 sorties

NOTES: The STV is a production-representative airframe with appropriate mass properties but without a guidance control unit or tail actuator subsystem. The GTV is a production JDAM guidance kit that may or may not have telemetry, a warhead, or a fuze.

RCS testing. Flight testing was limited to instrumented measurement vehicles to define the environment to be expected aboard each type of aircraft.

The purposes of EMD Phase 2 DT were to demonstrate aircraft compatibility, safe separation, maneuverability, accuracy, reliability, maintainability, supportability, and mission planning.

Because the existing general-purpose bombs to which JDAM is fitted did not change, LFT&E was not necessary.

AFOTEC as the lead OT agency conducted an MOT&E of the JDAM in conjunction with the Navy's OPTEVFOR. This MOT&E consisted of Air Force and Navy combined DT and OT, Air Force dedicated IOT&E with the B-52, and Navy OT-IIB OPEVAL on the F-18. All phases were structured to provide operationally realistic end-to-end mission scenarios, beginning with acceptance inspections and culminating with inert and live weapon drops. The B-52H and FA-18C/D delivery profiles were operationally representative, employing single and multiple weapons against single and multiple targets.

The Air Force and Navy conducted the combined DT and OT from July through October 1998 at the Naval Air Warfare Station China Lake, California, range complex. The Air Force conducted its dedicated IOT&E at the Utah Test and Training Range from November 1998 through July 1999. Aircrews from the 49th Test and Evaluation Squadron, operating from Barksdale AFB, Louisiana;

Andersen AFB, Guam; and Minot AFB, North Dakota, flew Air
Combat Command B-52Hs for these tests. The Navy conducted
OT-IIB at Naval Air Warfare Station China Lake, and on three
aircraft carriers from November 1998 through September 2000.

A total of 207 sorties were flown during the MOT&E phase. Of
these, B-52Hs flew 19 sorties (three captive-carriage and 16 weapon-
release sorties) and FA-18C/Ds flew 188(147 captive-carriage and 41
weapon-release sorties). All missions were planned using either the
AFMSS or the Navy TAMPS. The test team evaluated the effectiveness of these two systems, including AFMSS and TAMPS core; the
JDAM B-52H aircraft, weapon, and electronics; and FA-18C/D
mission-planning modules.

The test agencies rated JDAM as effective but not suitable
because of a combination of demonstrated weapon system performance against user requirements and the test team's judgment of mission accomplishment. JDAM met the user's requirements for captive-
carriage reliability for both threshold platforms. However, JDAM did
not meet the user's requirements for storage reliability, mission reliability, and system reliability. In particular, the guidance kit system
did not meet reliability thresholds, and the thresholds for mean time
to load and mission planning time were not met.

To address unresolved and unsatisfactory issues from IOT&E, a
dedicated FOT&E was planned. System reliability will be tracked
and evaluated through FOT&E and lot acceptance testing.

Joint Standoff Weapon

Mission
JSOW is a family of kinematically efficient 1,000-pound class air-to-
surface glide weapon.[7] It has LO and provides multiple kills per pass,
preplanned missions, standoff precision engagement, and launch-and-

---

[7] We derived this information from the JSOW Selected Acquisition Report of December 31, 1998;
DOT&E Annual Report FY 1996; Program Office (PMA 201) input via email dated April 5, 2002.

leave capability against a wide range of targets, day or night and in all weather conditions. JSOW is used for interdiction of soft or medium fixed, relocatable and mobile light and heavy armored targets, massed mobile armored targets, and antipersonnel and air-to-surface threats.

System Description

The JSOW Baseline (AGM-154A) consists of an airframe, a guidance-and-control system with INS-GPS capability, and a payload consisting of 145 BLU-97 submunitions. The JSOW/BLU-108 (AGM-154B) is similar, but the payload consists of 6 BLU-108 submunitions. The JSOW Unitary (AGM-154C) is a Navy-only variant that adds an autonomous imaging IR seeker to the INS-GPS guidance system and a BAe Broach warhead with penetrator capability. The all-up round is 160 inches long, has a wingspan of 106 inches when fully deployed, and weighs 1,065 pounds.

JSOW missions are normally preplanned using the Navy TAMPS or the AFMSS. JSOW interfaces with the aircraft through a MIL-STD 1553 data bus. The F-18C/D and E/F, F-16C/D, F-15E, JSF, B-1B, B-2A, and B-52H can employ this weapon.

Programmatics

The JSOW is a Navy-led, joint Navy–Air Force program. A DEM/VAL contract was awarded in June 1989. The Navy awarded the JSOW EMD contract in June 1992 to Texas Instruments Defense Systems and Electronics (currently Raytheon Missile Systems). The contract option for LRIP was exercised in February 1997. The Navy approved full-rate production for the AGM-154A in October 1998. In October 2002, the Air Force withdrew from the JSOW B program. The Navy completed development but deferred production.

Test Program

Table B.5 summarizes the test program, which the following paragraphs describe in more detail. Note that the description addresses only the JSOW Baseline.

Table B.5
JSOW Testing

Event	Guided Launches	Remarks
DT-IIA December 1994–March 1995	2	
DT-IIB March 1995–December 1995	10	At China Lake and Point Mugu
DT-IIC February 1996–October 1996	10	TECHEVAL and LFT&E
USAF DTE July 1996–August 1996	2	F-16 integration
OT IIA May 1996–September 1996	6	Operational assessment
OT-IIB February 1997–June 1997	14	OPEVAL Judged operationally effective and operationally suitable
IOT&E July 1998	2	OT for F-16

The Naval Air Warfare Center China Lake was the lead test activity. Other facilities used included Point Mugu, NAWC-AD (Patuxent River), and the Air Force Air Armament Center at Eglin AFB.

The Air Force began DT&E flight testing JSOW on the F-16 at Eglin AFB, Florida, in March 1996. AFOTEC conducted an operational assessment in December 1996, with 46th Test Wing conducting an update in April 1998 in conjunction with DT&E. The update focused on targeting and weapon platform performance, using the results of laboratory M&S, captive-flight, and free-flight testing. Less-than-desirable progress in integrating the baseline JSOW with the F-16 hindered Air Force testing. The problem was the control-section locking pins, a subassembly of the JSOW that Texas Instruments did not manufacture.

The U.S. Navy began OPEVAL testing in February 1997, after successful DT and initial OT programs. Over the entire test program, 42 of 46 shots were successful.

Sensor Fuzed Weapon

Mission

SFW delivers antiarmor munitions to neutralize a large number of enemy fighting vehicles in massed formations with a limited number of sorties.[8] It can be launched from a variety of altitudes and weather conditions and in countermeasure environments. Two related programs, the WCMD and JSOW, can be used with SFW to give it greater accuracy and range, respectively.

System Description

The SFW Cluster Bomb Unit (CBU-97/B) consists of three major parts: the tactical munitions dispenser, the BLU-108 submunition, and the projectile. The 1,000-pound-class tactical munitions dispenser holds ten parachute-stabilized BLU-108 submunitions. Each submunition holds four armor-penetrating projectiles, each of which has an IR sensor, for a total of 40 projectiles per SFW.

After the weapon is released, the tactical munitions dispenser opens and dispenses the ten submunitions. At a preset altitude sensed by a radar altimeter, a rocket motor fires to spin the submunition and initiate an ascent. The submunition then releases its four projectiles over the target area. The projectile's sensor detects a vehicle's IR signature, and an explosively formed penetrator fires at the heat source. If no target is detected after a specific time, the projectiles fire automatically, causing damage to material and personnel.

SFW is compatible with the A-10, F-15E, F-16, B-1, B-2 (WCMD only), and B-52 (WCMD only).

---

8 We derived the information in this section from the Sensor Fuzed Weapon Test and Evaluation Master Plan, August 1996, Area Attack Systems Program Office; Sensor Fuzed Weapon Test and Evaluation Master Plan, August 1999, Area Attack Systems Program Office; Information from Area Attack Systems Program Office (AAC/YH), Eglin AFB; Sensor Fuzed Weapon Selected Acquisition Report, December 31, 1998, SFW Operational Testing Summary, Director, Defense Operational Test and Evaluation; Number of launches from FY 1997 DOT&E Annual Report.

Programmatics

SFW began FSD in 1985. The development program was restructured in June 1989 because of test failures, schedule delays, and budget changes. The restructured program included a transition to production. After successful live-fire testing and a successful IOT&E from September 1990 to December 1991, OSD approved LRIP in March 1992. Decision authority was delegated to the Air Force in 1994, and the Air Force approved Milestone III for full-rate production in June 1996.

Two Producibility Enhancement Program (PEP) hardware upgrades were initiated for SFW to reduce costs and improve producibility through design improvement. The first, PEP-1, involved electronic and mechanical changes to the projectile, including the use of an application-specific integrated circuit that placed most of the electronic components on a single chip. The PEP 1 contract was awarded in early FY 1994. The PEP-2 contract was awarded in early FY 1995, and the program was to redesign the sequencer and altimeter into one integrated submunition electronics unit. The PEP-2 program was cancelled because of technical problems, and the critical elements were integrated into the P3I program.

In May 1996 the Air Force awarded a P3I contract for SFW. The P3I program involves three major improvements: (1) improving performance against countermeasures, (2) altering the warhead design to improve performance against softer targets without degrading the current target-set performance, and (3) raising the radar altimeter height of function to increase area coverage. The current sensor will be upgraded from a passive IR sensor only to a dual-mode type with passive IR and an active laser sensor. This upgrade will allow the sensors to discriminate between thermal and physical profiles of targets, enhance the sensor's performance against cooler targets, and improve its effectiveness against countermeasures. The improved warhead consists of a modified copper liner configured to form both a central penetrator for hard targets and multiple smaller penetrators for soft targets.

Test Program

Table B.6 summarizes the test program. DT included LFT&E using actual military vehicles instrumented for the test. Producibility modifications have necessitated several QT&E phases as well. FOT&E of PEP-1 was completed in 1998. All objectives were met, and testing results indicated that PEP-1 changes have not degraded the performance of the SFW.

Standoff Land-Attack Missile–Expanded Response

Mission

SLAM-ER provides standoff all-weather precision strike from carrier-deployed aircraft against fixed, high-value land targets and, secondarily, against relocatable stationary land targets and ships.[9] The missile

Table B.6
SFW Testing

Event	Launches	Remarks
DT&E December 1988– March 1992	39	38 missions 14 inert and 25 live rounds
LFT&E June 1990– September 1990		20 statically aimed submunitions to test for lethality against target Immobilized all land combat vehicles
IOT&E I September 1990– January 1992	36	AFOTEC conducted 30 sorties Assessed multiple kills per pass, effectiveness, and reliability and supportability
IOT&E II June 1995– February 1996	14	AFOTEC conducted 10 sorties Used LRIP weapons Found to be operationally effective at low altitude
FOT&E of PEP January 1998– August 1998	12	Air Warfare Center conducted three sorties

---

[9] We derived the information in this section from an interview with SLAM-ER Program Office on July 13, 2001 and from the FY 2000 DOT&E report for SLAM-ER.

is intended to fill the gap between long-range cruise missiles and short-range free-fall munitions.

System Description

The original SLAM was based on the Harpoon antiship missile, to which it added a GPS-aided INS for midcourse guidance, a Maverick imaging IR sensor, and a Walleye data link for man-in-the-loop control. SLAM-ER is a major upgrade with greater range; reduced susceptibility to countermeasures; greater capability against hardened targets; an improved guidance navigation unit; and improved user interfaces for mission planning, launch, and control. The primary changes from the existing SLAM included the following:

1. a modified Tomahawk Block III warhead
2. an improved data link with greater range and jam resistance
3. an improved guidance set with integrated GPS-INS, a 1760 data bus interface, a multichannel GPS receiver
4. modified Tomahawk wings
5. an automated mission-planning system.

Programmatics

SLAM-ER continued the SLAM development approach of maximizing use of existing components. After cancellation of the TASSM program, Boeing received an EMD contract in March 1995 to improve identified deficiencies of the interim SLAM. The first flight was in March 1997. LRIP I was approved April 1997, LRIP II in April 1998, and LRIP III in August 1999. Full-rate production was approved in May 2000, with existing SLAMs to be upgraded to SLAM-ER configuration. The threshold platform was the F/A-18C/D.

Test Program

Table B.7 summarizes the test program. There were five DT and eight OT launches. LFT&E was required because of the use of a new titanium-cased warhead. LFT&E consisted of confined volume test-

Table B.7
SLAM-ER Testing

Event	Guided Launches	Remarks
DT-1 March 1997	1	First flight Verified basic performance, mission planning, and maneuvering
DT-2 October 1997	1	Verified terrain following, data link, pilot designation, and target impact
DT-3 December 1997	1	Verified range for low-level launch and flyout, shift from IR track to designated aimpoint, and target impact
DT-4 February 1998	1	Verified range for high-altitude launch, off-axis launch, steep impact angle, and stop-motion aimpoint update
DT-5 March 1998	1	Verified new operational flight program, quick-reaction launch and midcourse update, target ID, and autonomous lock on a moving ship target
DT/OT-1 June 1998	1	Verified land-based target-of-opportunity capability
DT/OT-2 June 1998	1	Demonstrated the ability to attack a hardened aircraft shelter
DT/OT-3 June 1998	1	Demonstrated the ability to attack a high-altitude land-based target
DT/OT-4 September 1998	1	Consisted of production verification test
OT-IIA (OPEVAL) December 1998– May 1999	6	Six missiles fired in operationally realistic scenarios One prior DT/OT launch included in OT analysis Missile judged not suitable and not effective
VCD June 1999	1	Verified software and hardware (missile and AN/AWW-13 data link pod) changes
DT-II October 1999	1	
OT-IIB November 1999– January 2000	4	Four missiles fired in operationally realistic scenarios Deficiencies corrected Missile judged suitable and effective
ATA OT February 2002	1	Evaluate ATA capability

ing, three arena tests of warhead fragmentation, and four sled tests of warhead penetration.

Wind-Corrected Munition Dispenser

Mission

WCMD is an all-weather guidance kit that replaces the tail on the SFW (CBU-97), Combined Effects Munitions (CBU-87), and the Gator Mine System (CBU-89).[10] With the WCMD tail kit attached, these weapon designations become CBU-105, CBU-103, and CBU-104, respectively. WCMD is not a precision capability but does improve accuracy, depending on the quality of the navigation unit used. It can be launched from a variety of altitudes, weather conditions, and in countermeasure environments. WCMD corrects for wind effects, ballistic dispersion, and launch transients from moderate to high altitudes.

System Description

The WCMD kit integrates a low-cost INS, control unit, and steerable fins to guide tactical munition dispenser weapons. It is a bomb-on-coordinates weapon and is used by aircraft that have GPS-quality heading, velocity, and position data. The data are passed from the aircraft to the weapon via an MIL-STD-1760 interface. The threshold aircraft for compatibility are the F-16 and B-52. The objective aircraft are the B-1, F-15E, A-10, F-117, and JSF.

Programmatics

WCMD is an Air Force ACAT II program using commercial practices and government streamlining as much as possible. The contracting process minimized the use of military specifications. The acquisition emphasized affordability. Contractors received a performance requirement and price requirement for the average unit procurement and could trade-off performance objectives to minimize

---

[10] We derived the information in this section from the Wind Corrected Munitions Dispenser, Test and Evaluation Master Plan, Milestone III Revision, January 2001; the Wind Corrected Munitions Dispenser (WCMD) Engineering and Manufacturing Development (EMD) Cost Estimate, January 2001; and discussions with the Area Attack Systems Program Office (AAC/YH).

costs. The resulting average unit procurement cost for 40,000 units was approximately $10,000 in FY 1994 dollars.

The development program had two phases. EMD Phase 1 contracts for development with an option for pilot production were awarded in January 1995 to Alliant Techsystems and Lockheed Martin. The contractors built production-representative tail kits, and the Air Force held a fly-off competition. In January 1997, the Air Force selected Lockheed to continue into Phase 2, pilot production. In fall 1997, flight testing revealed an actuator problem during high-speed releases. The program was restructured to extend EMD with a third LRIP lot to allow time to design and qualify a fin-locking mechanism. In April 2001, WCMD was approved for full-rate production.

Test Program

Table B.8 summarizes the test program, which we describe in more detail below.

Table B.8
WCMD Testing

Event	Guided Launches	Number of Flights	Remarks
Contractor pre–fly-off tests January 1995– October 1996	11	10	Light testing verified predictions of contractors' six-degrees-of-freedom models
Fly-off November 1996– December 1996	14	8	
DT/OT	61	45	Approximately 60 captive-carry flights (approximately 120–180 flight hours)
Phase 1 May 1998– October 1998			

Table B.8—continued

Event	Guided Launches	Number of Flights	Remarks
Phase 2 October 1999– December 2000			
IOT&E January– October 2000	21	6	

The contractor used a digital system model capable of simulating INS characteristics, vehicle aerodynamics, actuators, inertial measurement units, and weapon operational flight program functions in various flight conditions, including launch and terminal maneuver.[11] Weapon simulator units were used to support lab testing to integrate the weapon to host aircraft.

Alliant and Lockheed conducted flight tests before the fly-off, using eight and nine test units, respectively. During the fly-off, F-16s dropped seven weapons from each contractor at Eglin AFB under various conditions.

The SPO and AFOTEC worked together to structure a combined DT/OT program for EMD after the fly-off. The test program had two phases. Phase 1 DT/OT involved the F-16 and B-52 and low dynamic pressure conditions; dedicated B-52 IOT&E followed. Phase 2 DT/OT and dedicated IOT&E involved the B-52 and F-16 and used pin-puller-configured hardware. The aircraft dropped examples of all three weapon types from various altitudes. The test program verified compliance with all the ORD key performance parameters, and no deficiencies are unresolved. The 46th Test Wing at Eglin AFB conducted the F-16 testing, and the 419th Flight Test Squadron at Edwards AFB conducted the B-52 testing. LFT&E was not necessary because WCMD does not modify the explosive characteristics of the munitions.

---

[11] The program office estimates that simulation eliminated at least 12 flights and drops.

System Test and Evaluation Work Breakdown Structure

This appendix presents an extract from the January 2, 1998, edition of MIL-HDBK-881, System Test and Evaluation.

H.3.3 System Test and Evaluation

The use of prototype, production, or specifically fabricated hardware/software to obtain or validate engineering data on the performance of the system during the development phase (normally funded from RDT&E) of the program.

Includes:

- detailed planning, conduct, support, data reduction and reports (excluding the Contract Data Requirements List data) from such testing, and all hardware/software items which are consumed or planned to be consumed in the conduct of such testing

- all effort associated with the design and production of models, specimens, fixtures, and instrumentation in support of the system level test program

> NOTE: Test articles which are complete units (i.e., functionally configured as required by specifications) are excluded from this work breakdown structure element.

Excludes:

- all formal and informal testing up through the subsystem level which can be associated with the hardware/software element
- acceptance testing

NOTE: These excluded efforts are to be included with the appropriate hardware or software elements.

H.3.3.1 Development Test and Evaluation

This effort is planned, conducted, and monitored by the developing agency of the DoD component. It includes test and evaluation conducted to:

- demonstrate that the engineering design and development process is complete.

- demonstrate that the design risks have been minimized.

- demonstrate that the system will meet specifications.

- estimate the system's military utility when introduced.

- determine whether the engineering design is supportable (practical, maintainable, safe, etc.) for operational use. provide test data with which to examine and evaluate trade-offs against specification requirements, life cycle cost, and schedule.

- perform the logistics testing efforts to evaluate the achievement of supportability goals, the adequacy of the support package for the system (e.g., deliverable maintenance tools, test equipment, technical publications, maintenance instructions, and personnel skills and training requirements, etc.).

Includes, for example:

- all contractor in-house effort

- (all programs, where applicable) models, tests and associated simulations such as wind tunnel, static, drop, and fatigue; integration ground tests; test bed aircraft and associated support; qualification test and evaluation, development flight test, test instrumentation, environmental tests, ballistics, radiological, range and accuracy demonstrations, test facility operations, test equipment (including its support equipment), chase and calibrated pacer aircraft and support thereto, and logistics testing

- (for aircraft) avionics integration test composed of the following:

  - test bench/laboratory, including design, acquisition, and installation of basic computers and test equipments which will provide an ability to simulate in the laboratory the operational environment of the avionics system/subsystem

  - air vehicle equipment, consisting of the avionics and/or other air vehicle subsystem modules which are required by the bench/lab or flying test bed in order to provide a compatible airframe avionics system/subsystem for evaluation purposes

  - flying test bed, including requirements analysis, design of modifications, lease or purchase of test bed aircraft, modification of aircraft, installation of avionics equipment and instrumentation, and checkout of an existing aircraft used essentially as a flying avionics laboratory

  - avionics test program, consisting of the effort required to develop test plans/procedures, conduct tests, and analyze hardware and software test results to verify the avionics equipments' operational capability and compatibility as an integrated air vehicle subsystem

  - software, referring to the effort required to design, code, de-bug, and document software programs necessary to direct the avionics integration test

  - (for engines) engine military qualification tests and engine preliminary flight rating tests

  - (for ships) model basin, hydrostatic, fatigue, shock, special sea tests and trials, etc., including the Extended Ship

Work Breakdown Structure (ESWBS), Trials Agenda Preparation, Data Collection & Analysis (842); Dock and Sea Trials (9823); and Hull Vibration Survey (9825) elements

H.3.3.2 Operational Test and Evaluation

The test and evaluation conducted by agencies other than the developing command to assess the prospective system's military utility, operational effectiveness, operational suitability, logistics supportability (including compatibility, inter-operability, reliability, maintainability, logistic requirements, etc.), cost of ownership, and need for any modifications.

Includes, for example:

- Initial operational test and evaluation conducted during the development of a weapon system

- such tests as system demonstration, flight tests, sea trials, mobility demonstrations, on-orbit tests, spin demonstration, stability tests, qualification operational test and evaluation, etc., and support thereto, required to prove the operational capability of the deliverable system

- contractor support (e.g., technical assistance, maintenance, labor, material, etc.) consumed during this phase of testing

- logistics testing efforts to evaluate the achievement of supportability goals and the adequacy of the support for the system (e.g., deliverable maintenance tools, test equipment, technical publications, maintenance instructions, personnel skills and training requirements, and software support facility/environment elements)

H.3.3.3 Mock-ups

The design engineering and production of system or subsystem mock-ups which have special contractual or engineering significance, or which are not required solely for the conduct of one of the above elements of testing.

H.3.3.4 Test and Evaluation Support

The support elements necessary to operate and maintain, during test and evaluation, systems and subsystems which are not consumed during the testing phase and are not allocated to a specific phase of testing.

Includes, for example:

- repairable spares, repair of reparables, repair parts, warehousing and distribution of spares and repair parts, test and support equipment, test bed vehicles, drones, surveillance aircraft, tracking vessels, contractor technical support, etc.

Excludes:

- operational and maintenance personnel, consumables, special fixtures, special instrumentation, etc., which are utilized and/or consumed in a single element of testing and which should be included under that element of testing

H.3.3.5 Test Facilities

The special test facilities required for performance of the various developmental tests necessary to prove the design and reliability of the system or subsystem.

Includes, for example:

- test tank test fixtures, propulsion test fixtures, white rooms, test chambers, etc.

Excludes:

- brick and mortar-type facilities identified as industrial facilities

Bibliography

Bell, Don G., *Naval Weapons Center Test and Evaluation Model (for Air-Launched Weapons)*, China Lake, Calif.: Naval Weapons Center, October 1988.

The Boeing Company, "Boeing 777 Facts." Online at http://www.boeing.com/commercial/777family/pf/pf_facts.html (as of January 28, 2004).

Cook, Cynthia, and John C. Graser, *Military Airframe Acquisition Costs: The Effects of Lean Manufacturing*, Santa Monica, Calif.: RAND Corporation, MR-1325-AF, 2001.

Commander Operational Test and Evaluation Force, *Operational Test Director's Guide*, COMOPTEVFORINST 3960.1H, December 13, 1995.

Defense Science Board, *Report of the Defense Science Board Task Force on Test and Evaluation*, Washington, D.C.: Office of the Under Secretary of Defense for Acquisition and Technology, September 1999.

Defense Systems Management College, *Test and Evaluation Management Guide*, 4th edition, Fort Belvoir, Va.: The Defense Acquisition University Press, November 2001.

Director, Operational Test and Evaluation, Annual Report to Congress for Fiscal Year 2000.

DoD—see U.S. Department of Defense.

F-22 System Program Office, *F-22 Test & Evaluation Master Plan*, Wright-Patterson AFB, Ohio, June 1999.

Gogerty, David C., Bruce M. Miller, J. Richard Nelson, Paul R. Palmer, Jr., *Acquisition of Contemporary Tactical Munitions*, Vol. I: *Summary Report*, Alexandria, Va.: Institute for Defense Analyses, March 1990.

Jonson, Nick, "Merits of Streamlining DoD Test and Evaluation Operations Debated," *Aerospace Daily*, May 22, 2002.

Lorell, Mark, and John C. Graser, *An Overview of Acquisition Reform Cost Savings Estimates*, Santa Monica, Calif.: RAND Corporation, MR-1329-AF, 2001.

U.S. Air Force, *Developmental Test and Evaluation*, Air Force Instruction 99-101, November 1, 1996.

_____, *Operational Test and Evaluation*, Air Force Instruction 99-102, July 1, 1998.

_____, *Armament/Munitions Test Process Direction and Methodology for Testing*, Air Force Manual 99-104, August 1, 1995.

_____, *Live Fire Test and Evaluation*, Air Force Instruction 99-105, July 25, 1994.

_____, *Test Resource Planning*, Air Force Instruction 99-109, July 21, 1994.

_____, *Test and Evaluation: Airframe-Propulsion-Avionics Test and Evaluation Process*, Air Force Manual 99-110, July 3, 1995.

U.S. Air Force Scientific Advisory Board, *Report on Review of the Air Force Test and Evaluation Infrastructure*, SAB-TR-97-03, November 1998.

U.S. Department of Defense, Weight and Balance Data Reporting Forms for Aircraft (Including Rotorcraft), MIL-STD-1374, September 13, 1977.

_____, *Department of Defense Handbook: Work Breakdown Structure*, Appendix H, MIL-HDBK-881, January 2, 1998.

_____, *Major Range and Test Facility Base (MRTFB)*, DoD Directive 3200.11, January 26, 1998.

_____, *Mandatory Procedures for Major Defense Acquisition Programs (MDAPS) and Major Automated Information System (MAIS) Acquisition Programs*, DoD 5000.2-R, April 5, 2002.

U.S. General Accounting Office, *Best Practices: A More Constructive Test Approach Is Key to Better Weapon System Outcomes*, GAO/NSIAD-00-199, July 2000.

Younossi, Obaid, Mark V. Arena, Richard M. Moore, Mark Lorell, Joanna Mason, and John C. Graser, *Military Jet Engine Acquisition: Technology Basics and Cost-Estimating Methodology*, Santa Monica, Calif.: RAND Corporation, MR-1596-AF, 2003.

Younossi, Obaid, Michael Kennedy, and John C. Graser, *Military Airframe Costs: The Effects of Advanced Materials and Manufacturing Processes*, Santa Monica, Calif.: RAND Corporation, MR-1370-AF, 2001.